Life Explained

Published in association with Éditions Odile Jacob for the purpose of bringing new and innovative books to English-language readers. The goals of Éditions Odile Jacob are to improve our understanding of society, the discussions that shape it, and the scientific discoveries that alter its vision, and thus contribute to and enrich the current debate of ideas.

MICHEL MORANGE, *Life Explained* (2008)

Life Explained

Translated by Matthew Cobb
and Malcolm DeBevoise

MICHEL MORANGE

Yale University Press New Haven and London

Éditions Odile Jacob Paris

Odile
Jacob

A Caravan book. For more information, visit www.caravanbooks.org.

Translated from *La vie expliquée? 50 ans après la double hélice,* by Michel Morange, published by Éditions Odile Jacob in 2003. Copyright Odile Jacob, 2003; ISBN 2-7381-1282-X.

Set in Minion type by Integrated Publishing Solutions.
Printed in the United States of America.

Library of Congress Cataloging-in-Publication Data
Morange, Michel.
[Vie expliquée. English]
Life explained / Michel Morange ; translated by Matthew Cobb and Malcolm DeBevoise.
p. cm.
ISBN 978-0-300-13732-3 (alk. paper)
1. Biology—Philosophy. I. Title.
QH331.M6313 2008
570—dc22 2008017943

A catalogue record for this book is available from the British Library.

This paper meets the requirements of ANSI/NISO Z39.48-1992 (Permanence of Paper). It contains 30 percent postconsumer waste and is certified by the Forest Stewardship Council.

10 9 8 7 6 5 4 3 2 1

Life is a term, none more familiar. Any one almost would take it for an affront to be asked what he meant by it. And yet if it comes in question, whether a plant that lies ready formed in the seed have life; whether the embryo in an egg before incubation, or a man in a swoon without sense or motion, be alive or no; it is easy to perceive that a clear, distinct, settled idea does not always accompany the use of so known a word as that of life is.

—JOHN LOCKE, *An Essay on Human Understanding*

It is no doubt a very important matter to enquire into the nature of what is called life in a body. . . . However difficult may be this great inquiry, the difficulties are not insuperable; for in all this we have to deal only with purely physical phenomena.

—JEAN-BAPTISTE LAMARCK, *Zoological Philosophy*

Contents

Preface

More than half a century ago, in 1953, Francis Crick and James Watson discovered the double helix structure of the DNA molecule, which constitutes the basis of heredity. They thought, Crick said, that they had discovered "the secret of life."

In the quarter century between 1940 and 1965, scientific understanding of the fundamental phenomena of life made remarkable advances. But had the secret of life therefore been discovered? Had the question "What is life?" really been answered? Many people believed—and still believe today—that it had. Scientists, in particular, scarcely bothered any longer to inquire into the nature of life. By the latter part of the twentieth century the question had become old-fashioned, even taboo, since to ask "What is life?" would have implied that the answer already given by molecular biology was somehow not acceptable. By posing the question, one risked being excluded from the mainstream of modern science.

Things have begun to change, however. Fewer and fewer scientists are convinced that we have the complete answer. The question has once again become respectable, and it now lies at the heart of research being carried out by a great many biologists and other scientists in a wide range of fields. More than fifty years after the publication of Erwin Schrödinger's famous essay *What Is Life?* (1944), it has begun to reappear as a book title, and regularly occurs in the opening lines of journal articles on the origin of life.

The epigraph from John Locke at the beginning of this book is meant to call attention to two outstanding aspects of the question "What is life?": first, its antiquity, and the large number of answers that have already been given to it; and second, the historical character of the question itself, which three hundred years ago was posed in a very different way than it is today. Although the use of the term "life" is not obviously any clearer now than it was in Locke's time, the reasons for this are not necessarily the same. Today, for example, we unequivocally accept that both a plant seed and an unincubated hen's egg are alive. As the French philosopher of science Georges Canguilhem pointed out, both the nature of the answer given to the question "What is life?" and the interest shown in these answers changed considerably during the centuries following Locke. In this respect, the twentieth century was no different from the eighteenth and nineteenth.

My chief purpose is to extend this history beyond Watson and Crick's discovery of the double helix—the point reached by Canguilhem—and to describe some of the new ideas that have been stimulated by developments in biology over the past half century. Many of these developments have gone unnoticed by the general public, as though it were supposed that the need for research had come to an end. The revival of the question "What is life?" is a consequence also of work taking place outside biology proper, notably in the emerging field of astrobiology, as part of a more general change in intellectual orientation that I will examine as well.

Although the question is once again respectable, even fashionable, only a few scientists have dared to try to answer it directly. Their replies, though inevitably different from ones that have been given in the past, nonetheless form the latest chapter in a rich philosophical and scientific tradition. In addition to these explicit answers, others have implicitly been proposed by new research in biology and related fields. The attempt to create a functional biological membrane, for example, embodies a different conception of the steps involved in the formation of life than the attempt to create self-replicating nucleic acid polymers in a test tube; taken together, they represent two distinct ways of characterizing life. Experiments performed on a space probe in the hope of detecting the presence of life on other planets are another way of implicitly characterizing life.

Any increase in our knowledge of organisms and life unavoidably has implications for the problem of life's origins and nature. This may be why most of the great biologists have studied it at one time or another. The career of the French molecular biologist and Nobel Prize winner Jacques Monod is a case in point. For many years Monod showed no interest in the origins of life; but when he started work on *Chance and Necessity* (1970), which contains a systematic discussion of the philosophical and ethical impact of recent advances in the molecular understanding of life, he found himself increasingly preoccupied by this question.

On surveying the various answers that have been given to the question "What is life?" a consensus can be seen to emerge with regard to the fundamental characteristics that are shared by organisms, and that therefore jointly constitute life. This consensus rapidly disappears, however, when scientists try to classify these characteristics and rank them in a hierarchy with a view to isolating a single characteristic that is consubstantial with life. Nonetheless we can all agree with Lamarck—the source of my second epigraph—that inquiring into the nature of life from a scientific point of view is indeed a very important matter. Another part of my purpose in writing this book, then, is to help readers think about questions that are seldom dealt with directly in science, and to try to reforge the bonds that once united science and philosophy (or what may be called, more generally, the humanities).

In seeking to act as an intermediary between biological research and philosophical thought, I wish to acquaint philosophers (and historians of ideas) with new discoveries and concepts in biology, enabling them to go beyond the DNA double helix and the informational conception of life that for the moment, at least, appears to have triumphed over other views; and to introduce biologists to the rich philosophical (and historical) tradition of thinking about life. I am well aware that this is a thankless task, for it cannot help but invite attacks from both sides. No doubt I will be accused, on the one hand, of having failed to present the latest scientific research in sufficient detail, and, on the other, of caricaturing the philosophical issues. But I would rather risk such reproaches than accept the present situation, in which philosophers argue only with dead biologists and biologists only with dead philosophers, while counting on each side to correct my mistakes and make up for my omissions.

Some may ask why one should seek to reconcile science and philosophy—what the British scientist and novelist C. P. Snow famously called "the two cultures." It could be argued, after all, that serious scientists would do well simply to ignore a vague question like "What is life?" because the time and energy involved in wrestling with it are wasted: whatever answer may be given, it will not have any immediate or direct effect on the actual course of research. I will return to this objection later on, and consider it in greater detail, for it is not wholly without merit. But were scientists to accept it completely, they would end up consigning themselves to an intellectual ghetto. Science has drawn—and continues to draw—great strength and vitality from the attempt to answer the great questions that humanity has posed itself throughout the ages. If scientists were to sever their historical connection with these sources of inspiration, the future development of science is likely to be severely hampered, and its quality compromised.

This book cannot help but bear the mark of my own training. I adopt the point of view of a molecular biologist, because that is what I am. A theoretical biologist or a researcher into the origins of life would no doubt have chosen different data and presented them differently. Nonetheless, I trust that I have cited the contributions of other disciplines often enough that interested readers will be able to consult them and, in the event of disagreement, to make up their own minds.

I should say, too, that my approach differs from that of most recent authors on the subject, many of whom are deeply involved in research into the origins of life, and who attempt to show that their view of the matter is superior to the ones that have been given by their competitors or by earlier writers. It is no doubt a weakness that I am less well informed about the technical aspects of the various models currently being debated than they are. But I hope that it is also a strength. I am less likely to be swayed by doubtful evidence, more apt to be skeptical of claims cloaked in fine phrases that turn out on closer examination to have no precise meaning. Nor do I have any personal or professional interest in advocating one line of research rather than another. For all these reasons I believe I am in a better position to provide an objective—or at least a neutral—account of the various approaches.

Readers of this book will therefore not find the sort of detailed technical discussion—for instance, of the synthesis of polymers in prebiotic media—that can readily be found in other works (which I nonetheless mention, and to which they are free to refer). I wish instead to present the recent evolution of ideas in the historical context of discussions about life, in order to be able to appreciate the weight and worth of current views; to distinguish what is genuinely new from what amounts to nothing more than an unwitting revival of old ideas. And although I readily acknowledge the great importance of research on the origins of life in this connection, many other lines of inquiry (to say nothing of social and cultural transformations) have contributed to the development of recent thinking, as I shall try to show in the pages that follow.

It is a pleasure to thank all those who, directly or indirectly, have supported me in this project. Anne Fagot-Largeault read the original manuscript with an attentive and critical eye, helping me both to sharpen the argument and, at least in part, to assuage my fears about venturing into the realm of philosophy. The encouragement and comments of the ecologist Régis Ferrière, a close colleague, as well as those of the philosopher Lindley Darden, were extremely valuable also. I owe thanks, too, to my translators, Matthew Cobb, who made a preliminary English version of the original French text, and Malcolm DeBevoise, who extensively revised the draft translation and made a number of useful editorial suggestions. Finally, I am indebted to my French publisher, Odile Jacob, and especially to Yale University Press, whose referees helped me improve the final manuscript still further.

Part I
The Death and Resurrection of Life

In 1962, a French biologist named Ernest Kahane published a book titled *Life Does Not Exist.* Much more recently an American, Stanley Shostak, published one called *Death of Life.* Although both authors described the same phenomenon—the abandonment by biologists of research into the nature of life—their accounts of it were very different. For Kahane, the question of life, having been reduced to a physicochemical problem, had therefore either been solved or was in the process of being solved. Shostak, on the other hand, to judge from the subtitle of his book ("The Legacy of Molecular Biology"), felt that the problem of life had not so much disappeared as it had been overshadowed by a new picture of the living world. Indeed, only a few years after the appearance of Kahane's book, the French molecular biologist and Nobel Prize winner François Jacob, in *The Logic of Life* (1970), remarked that life was simply no longer a topic for discussion in biology laboratories.[1]

Between the 1960s and the 1990s, the question "What is life?" virtually disappeared from scientific discourse. For most biologists its disappearance signified the definitive rejection of all forms of vitalism.[2] Toward the end of the twentieth century, merely to pose the question carried the risk of exposing oneself to ridicule as a spiritualist of one kind or another, for whom life was something more than the product of particularly complicated chemistry. A second reason why the question was no longer asked is that many biologists felt that a clear answer had in fact been provided by molecular biology. That the question should now have come back into fashion—that the corpse of life seems still to be twitching—does not mean that spiritualism is making a comeback, but that previous scientific answers to the question are no longer considered fully satisfactory. Its revival is therefore due to the evolution of biological ideas. By themselves, however, developments within biology would not have been enough. Outside influences played a major role as well.

Life passed into eclipse only for a few decades—a brief and exceptional interlude in the long history of the biological sciences. For biology cannot escape a question that lies at its very heart.

1
The Twilight of Life

The expression "twilight of life" was first used in 1935, in an article that appeared in the *New York Times*, to summarize the implications of the crystallization of the tobacco mosaic virus (TMV) by the American chemist Wendell Meredith Stanley.[1] This experiment showed that the TMV was no different from the molecules that were habitually manipulated and purified by organic chemists: it was merely a very large molecule.

Stanley's experiment was spectacular, and it received an enormous amount of publicity. But it was only one of a number of experimental approaches that, by the middle of the twentieth century, had succeeded in depriving organisms of their mystery and substituting in its place the chemistry of macromolecules. Thirty years later a sequel to Stanley's discovery attracted nearly as much media attention. In 1967, using only simple molecules, the American biochemist Arthur Kornberg managed to replicate the genetic material of a bacteriophage—a small virus that infects bacteria—in the test tube ("in vitro") by introducing a specific enzyme.[2]

To understand the philosophical import of these experiments, we need to put them in the context of a long historical debate. From the seventeenth century onward, naturalists had sought to describe the function of organisms in terms of physical principles. The first mechanistic models failed to withstand the decisive test of experiment, however. The development of physiology in the middle of the eighteenth century, and the invention of the term "biology" at the beginning of the nineteenth century, were clear signs of a vitalist reaction to the simplistic reductionism of these models.[3]

Chemists nonetheless slowly learned first how to isolate the molecules of life, and then how to make them. Although the road that led from the synthesis of urea by Friedrich Wöhler in 1828 to the in-vitro fermentation of sugars by Eduard Büchner in 1897 was hardly straight, it pointed in a single direction. A few years later, another German chemist, named Wolfgang Ostwald, used the phrase "world of neglected dimensions" to describe the terra incognita that lay between the molecules studied by organic chemists and the complex internal structures of cells that could barely be discerned under a light microscope.[4] It was this world that biologists forcibly invaded in the opening decades of the twentieth century. The characterization of metabolic pathways and the plodding, but constant, progress in the description of biological macromolecules all followed from the work of Wöhler and Büchner, which, in seeking to naturalize the functioning of organisms, began to pull down the barrier that separated the chemistry of life from that of the inanimate world. From this point of view, the crystallization of the TMV was just another step forward. But the TMV was not a mere macromolecule: it was a virus, and therefore, it was argued, an organism. Even if it was very simple, it was nevertheless on the animate side of the boundary between life and non-life. By crystallizing the TMV, Stanley had crossed this boundary and, at the same time, destroyed it.

In the early decades of the twentieth century, viruses occupied an increasingly important place in biological research. By the late nineteenth century they were considered an extremely small kind of microbe, because they passed through filters that retained other microbes. Impossible to grow in vitro, they were known to be responsible for serious pathologies in humans (influenza and polio, for example) and also for diseases in animals and plants. But while viruses attracted particular attention on account of their medical and economic interest, they were studied mainly because they were seen to be an elementary form of life.[5] What is more, because of their small size and simple chemical composition, they appeared to be within reach of the most advanced physicochemical techniques. Some researchers (the Canadian bacteriologist Félix d'Herelle, for example) even thought that they were fossil traces of the first forms of life that had appeared on the planet.

The widespread interest in viruses also grew out of their resemblance to genes. Although Gregor Mendel had discovered the laws of genetics in 1865, he had not given a name to the "thing" that enabled characters to be transmitted, nor had he suggested what it might be made of. The reification of the gene, as it might be called—its transformation into an object that could be studied using physical and chemical tools—did not begin until 1910, when Thomas Hunt Morgan's group at Columbia University demonstrated that genes are linked to chromosomes.[6] What made the gene interesting was not only its role in determining characters, but also its capacity for self-replication, which seemed analogous to the capacity for reproduction observed in organisms. At the same time, the gene, like the virus, was sufficiently small that its physicochemical properties could be studied.[7] As the smallest "unit of life," it lay at the intersection of research by biologists on the smallest possible hereditary units within organisms and inquiry by physicists into the structure of the most complex possible molecules.[8] It was therefore the ideal research object, possessing the fundamental properties of life (self-replication and variation) in their simplest form. In 1929, the American geneticist Hermann Muller put forward the hypothesis that genes form the basis of life itself.[9]

The realization that viruses and genes shared a number of salient characteristics—their ability to replicate themselves with variation, their small size, and what was assumed to be their critical role in the earliest phases of life—had already led Muller, in 1922, to propose that viruses (in particular, bacteriophages) were pure genes.[10] This identification gave further impetus to the study of viruses, and lent even greater importance not only to Stanley's experiment on the TMV, but also to research on the bacteriophage being carried out at about the same time by the German-born physicist Max Delbrück.[11] There is a striking contrast, however, between the impact of Stanley's experiment, which helped to expel the last vestiges of vitalism from biology, and the discreet and almost simultaneous movement away from the idea that viruses were a suitable model for the study of organisms. Viruses, it gradually became clear, are obligatory parasites—simple forms of life that use the machinery of host organisms to reproduce themselves. The decline of the virus model occurred in stages. The first doubts were raised in the mid-1930s, at the same time that Stanley succeeded in isolating the tobacco mosaic virus.[12] The problem was that, despite a great many attempts, it proved impossible to cultivate viruses in any non-living medium. With greater insight into the fundamental molecular mechanisms of life came a better understanding of the reasons for the strict parasitism of viruses, which were discovered to be nothing more than packets of genetic information protected by a more or less complex envelope of proteins. Viruses have neither the necessary molecular machinery to read this information nor a metabolism capable of constructing such machinery.

The "twilight of life"—the widespread expectation that life's mystery would finally be dispelled with the unlocking of its secrets—was thus in fact a twilight only for objects that, because they are not autonomous and do not have the extraordinary ability to synthesize chemicals, lack the distinctive characteristics of life. This paradox evaporates, however, if we introduce a distinction between "replication" and "reproduction" that does not arise in the common use of these two terms, and that has been obscured further in recent decades by the genocentric view of the living world promoted by the British ethologist and evolutionary biologist Richard Dawkins.[13] To replicate is to make a faithful copy of an object. Photocopies are a form of replication. The duplication of a DNA molecule into two daughter molecules is likewise a process of replication. On the other hand, reproduction in the biological sense implies the existence of a complex autonomous organism and its participation in the creation of a second organism that is similarly autonomous. The term "reproduction" therefore refers to a complex process involving entities with complex structures and functions.

In the case of both viruses and genes, only the term "replication" is appropriate; "reproduction" implies an autonomy that neither one possesses. Confusing these terms—and the distinct processes they describe—has had one very significant consequence, namely that the reproduction of organisms is often reduced to the replication of the molecules that form them. In retrospect we can see that the momentary importance of viruses in the explanation of organic phenomena was due to a "hard" form of reductionism that denies the possibility that characteristics or functions may require a certain degree of complexity in order to be expressed, and seeks to explain them instead by reference to the structure of one or a few elementary components. In reducing the complex phenomenon of reproduction to the mere replication of macromolecules, it went unnoticed that the concept of reproduction itself had been deformed and denatured.

The reduction of life to physicochemical phenomena has had the further consequence—a very important one—of favoring research into the use of organisms for commercial purposes. It is not by accident that the development of this field, biotechnology, should have coincided with the growing domination of a reductionist conception of life.[14]

2

Life as Genetic Information

In July 2002, the American journal *Science* announced the in-vitro synthesis of the polio virus (more precisely, the synthesis of a nucleic acid that allows the virus to be produced once it has been inserted into a cell) from simple molecules.[1] The news made front pages around the world.

This kind of media attention may seem rather surprising. After all, the experiment was not entirely novel. As we have already seen, similar studies of a bacterial virus had been made thirty-five years earlier. The excitement surrounding this announcement had to do instead with memories of the devastating effects of the polio virus; admiration for the technological advances in the interval that had made it possible to synthesize the virus using only information contained in data banks, without direct reference to an actual virus; and fears that terrorists, employing the same method, might succeed in once again spreading a virus that was gradually being eradicated through an ongoing and global campaign of vaccination.

But the most curious aspect of this affair (and what led me, in fact, to write this book) was the confused nature of the questions posed by both journalists and scientists regarding the study's implications. Had the experimenters actually *created* a polio virus? If so, had they created an organism? Was there any difference between their creation and that of the Creator? Is life inside a cell any different from life in a virus? The answers that were given to these questions turned out to be even more confused.

Already in the 1930s it had become apparent that viruses were not adequate models for illustrating the "principle" of life. And yet still today they are assumed, at least by the media, to satisfy this purpose, if only because the mere creation of a virus obliges its authors to deny that they have been playing God. The reason for this is that, beginning in the 1920s, viruses were taken to be genes, and the appearance of genes was thought to be identical with the appearance of life. Despite later developments, this genetic and informational view of life is still dominant in journalistic accounts of recent discoveries.

More than sixty years ago, in his famous essay on life and the origins of the order observed to exist in organisms, the Austrian physicist Erwin Schrödinger argued that this kind of order clearly differs from the one found in the inanimate world, which is based on statistical laws that account for the movement of particles.[2] Schrödinger predicted that the origins of order in the living world would be discovered in the specific molecular structure of those parts of the cell that seemed to be chiefly responsible for its function: genes and chromosomes. Chromosomes were understood to be the carriers of genetic information, which was transmitted from generation to generation, enabling both the structural and functional characteristics of organisms to be reproduced.

Over the next three decades, from the mid-1940s through the 1960s, research in molecular biology provided support for

Schrödinger's claims and allowed them to be precisely stated in chemical terms. The active agents in cells, as we now know, are proteins—macromolecules that act as catalysts, activating chemical reactions, receiving and transmitting molecular signals, and endowing cells with form and mobility. Proteins are formed by chaining together smaller molecules—amino acids—in a specific and predetermined sequence. This sequence is not directly transmitted from generation to generation; instead it is indirectly coded in another macromolecule, DNA. Decoding this sequence permits the synthesis of the proteins responsible for the incessant chemical transformations that take place inside the living cell, and for its reproduction. With the discovery of the simple double helix structure of DNA in 1953 by James Watson and Francis Crick, it became possible to understand the ease with which this molecule replicates itself, and also how the information needed for the precise synthesis of proteins could be contained in its nucleotides.

This new informational conception of life made it possible to understand the nature of viruses as well. Viruses generally consist of a single molecule of nucleic acid (RNA or DNA), with one or more protein envelopes that protect the genetic information during its passage from organism to organism.[3] It is this information that permits the synthesis of the enveloping proteins and of the few enzymes necessary for the replication of the genetic material. By itself a virus cannot replicate: it does not produce the energy required either for protein synthesis or for the replication of its own nucleic acid molecules; it is unable to manufacture the molecular components of its macromolecules; and it does not possess the highly complex molecular structures that allow genetic information, stored in the form of nucleic acids, to be translated into proteins. The structure of DNA and its role in coding the information needed for protein synthesis are now so familiar that we are apt to forget that the discovery of these fundamental molecular mechanisms of life came as a great surprise and source of wonder. Nothing in the many studies that have been carried out since has cast the least doubt upon the elegance and functional efficiency of these mechanisms, which have been retained by evolution to ensure the nearly exact reproduction of life forms.

It was tempting to suppose that the long-sought explanation of living phenomena had been found in the perfection of these very mechanisms—all the more so because they operate identically in all organisms. Little wonder, then, that by the 1960s Crick, Monod, and other molecular biologists were convinced that they had discovered the secret of life. To be sure, a great many details remained to be worked out, not least among them the mechanisms involved in cellular differentiation and the embryonic development of multicellular organisms, including the formation of highly complex structures such as the brain. But the constitutive principle of life had been identified: the genetic code, which is to say the rule of perfect correspondence between the structure of DNA (molecular memory) and that of proteins (the active agents of life and cellular form).

For several years this understanding was confined to the theoretical level. Then, in the early 1970s, molecular tools were developed that made it possible to modify genetic material for the purpose of altering the properties of organisms. Most experiments had the basic aim of clarifying the role of various genes in the development or the functioning of a given organism. But it was only a short step from the basic to the applied, from understanding a gene's function to modifying the genome in order to create a new type of organism. Bacteria were manipulated to produce animal or human proteins; plants were made resistant to various pathogenic agents, or to toxic substances such as weed killers; laboratory animals (especially mice) were genetically modified. The ability to modify an organism at will and to endow it with new properties (within the limits of the current understanding of gene function), and in this way to exert control over the evolution of life, was a striking indication of the progress that had been made in our understanding of the living world.

Furthermore, many people found it satisfying to imagine that organisms should contain a code and processes of translation analogous to the ones used by human beings in connection with what is widely believed to be their distinguishing characteristic: language. Some took this to be proof that human language had a direct biological origin; that, as the gospel according to John puts it, at the very beginning of the world—or, at least, at the beginning of life—was the word, speech. Life was *logos.*[4]

All these comparisons and analogies now look very naive. The fact that organisms contain a genetic code does not mean that they have a language in the sense that we use this term in everyday speech. And yet we should not be overly harsh in criticizing the unfounded assumptions of those who came before us. The discovery of the genetic code was a monumental shock, the like of which is rarely encountered in science and which readily excuses the handful of overstatements that were later made in its name.

Several other lines of research led to the idea that life could be reduced to non-life through the intermediation of macromolecules and information. For the moment I will mention only two of them, while reserving this topic for more detailed discussion later. The first involves two similar accounts of the origin of life that were developed virtually simultaneously in the 1920s by the Russian biochemist Aleksandr Oparin and the British biochemist and geneticist John Haldane.[5] The force of these accounts derived from the grouping together of data and observations from different disciplines, including biochemistry, geology, and astrophysics (the study of planets and their formation). Each of the scenarios devised by Oparin and Haldane assumed a series of three stages:

1. Molecules such as amino acids, which were later to serve as the building blocks of life, appeared in the primitive chemically reducing conditions thought to have existed on Earth in the early phases of its development.
2. These molecules then spontaneously associated with one another to form macromolecules.
3. Self-reproducing metabolic systems in chemical exchange with the external environment appeared in spontaneously formed, closed structures (called "coacervates" by Oparin).

Grounds for speculation of this sort were greatly strengthened by an experiment conducted in 1953 by Stanley L. Miller at the University of Chicago, under the supervision of Harold C. Urey, that reproduced the first stage in vitro. Miller showed that amino acids were formed from simple molecules such as ammonia and methane when these substances were held in a reducing atmosphere and subjected to repeated electrical discharges.[6] The key stage of the Oparin-Haldane hypothesis was soon seen to be the biosynthesis of macromolecules and, above all, the appearance of a functional relation between these macromolecules—the invention of the genetic code.

A second line of research had its origins in studies of information, the development of electronic calculating machines (later called computers), and the cybernetics movement, all of which flourished during and after World War II. There is no need to go into the details of this research to understand its profound impact on conceptions of life. The essential thing to keep in mind is that it had now become possible to imagine the development of machines that were sufficiently powerful to reproduce all the complexities of life. Progress toward this objective, through the construction of the first computers and the implementation of complex cybernetic networks, quite naturally led mathematicians like Alan Turing and John von Neumann to ask what, if anything, distinguished organisms from machines that could imitate the behavior of organisms. In particular, the question arose whether a machine could cross the boundary between the inanimate and the living worlds. It was in this context that von Neumann proposed his theory of automata, in a historic paper published in 1948 that sketched the outlines of a machine that could reproduce itself by locating the necessary components in its environment.[7]

In the absence of any precise idea about the physical and chemical composition of organisms, these studies of artificial life long remained purely theoretical. Nevertheless, they had considerable influence and contributed, at least in the popular mind, to a blurring of the boundary between life and non-life. They are still frequently cited today in technical discussions about the origin and nature of life.[8]

3

The Return of Life

Molecular biologists used to argue that the appearance of life and the appearance of genetic information were one and the same thing. The plausibility of this claim derived from the fact that all terrestrial organisms are descended from the same ancestor and use the same genetic code. But it was sophistry nonetheless, and did more to obscure the origin of life than to clarify it. The problem was not so much the first stages (the formation of molecules and macromolecules) as the appearance of precise rules of correspondence between these macromolecules. In particular, the molecular biologists' view failed to explain how the genetic code could have appeared in the first place, given that decoding requires cells to use proteins that are themselves coded for in the genome.

The work of the American scientist Carl Woese and of two British-born scientists working in the United States, Francis Crick and Leslie Orgel, soon reduced the apparent complexity of these functional organic macromolecules to just two components, RNA and proteins, by eliminating DNA (whose chemical properties suggested that it was a late arrival on the evolutionary scene).[1] Nevertheless, the fundamental problem remained. This led many molecular biologists, including Jacques Monod, to think that the appearance of life forms must be an extremely rare event in the history of the universe; that life as we know it is uniquely confined to Earth.

Beginning in the early 1980s, conventional wisdom found itself challenged again with the demonstration that RNA could also act as a catalyst—a role that previously had been reserved for proteins. In 1985, the American molecular biologists Norman Pace and Terry Marsh proposed the presence of a self-replicating RNA entity at the origin of life.[2] A year later another American, Walter Gilbert, put forward the hypothesis that the living world we know was preceded by an "RNA world," in which this macromolecule combined the functions that are now separately carried out by DNA and proteins. On this view, RNA acted both as memory and as functional agent, in particular as a catalyst.[3] The RNA world then gave rise to proteins, which together with RNA subsequently led to the appearance of the third macromolecule, DNA. This model is now generally accepted, despite difficulties and doubts that the original nucleic acid molecule was in fact RNA, which, because it is chemically very unstable, is unlikely to have appeared first.[4] In the wake of Gilbert's hypothesis a number of others were advanced, some of which had been proposed earlier but had failed to attract much attention. Some researchers suggested that the RNA world had been preceded by other living worlds in which replication processes were based on some other material factor, no longer serving this purpose, such as crystalline surfaces.[5]

The historical reality of such alternative living worlds is not universally admitted. The key point, however, is that these models opened the way to speculation about the nature of life and its origin by demonstrating the falsity of the earlier view, which identified life with genetic information in the form of a DNA code. It was now clear that genetic information—and the genetic code—were later inventions of the living world that enabled organisms to ensure their reproduction more effectively.

Once again, terminological ambiguity has led to a great deal of confusion. Biological macromolecules—DNA, RNA, and proteins—are often said to be "informational" (or "informative") macromolecules. This may only be a loose way of emphasizing their complexity, which arises from the fact that they are formed by linking elementary components together in a precise sequence; but it could also mean that macromolecules (mainly nucleic acids) contain the information required for the synthesis of other macromolecules.[6] A macromolecule can induce the formation of another macromolecule identical to itself (as in the case of an RNA molecule, which can self-replicate directly or indirectly), or it can induce the formation of a different type of macromolecule (thus the sequence of amino acids that form a protein is coded in present-day organisms by a gene made of nucleic acid). It is obvious that informational macromolecules in the first sense of the term appeared very early in the evolution of life: it was their structural characteristics that enabled the first cells to function. The difficulty is that if we adopt this sense of "informational," everything is informational, since everything has a certain degree of complexity. The appearance of informational molecules in the second sense—as the carriers of information governing the synthesis of similar or different macromolecules—probably occurred much later, however.

The term "genetic" suffers from the same ambiguity. In present-day life forms, genetic material (that is, the material that enables characters to be transmitted from cell to cell and from generation to generation) is made up of nucleic acids. But a nucleic acid is not, in and of itself, a genetic polymer, even if its structure makes it possible for it to become one. It will become a genetic polymer only if it is placed in a network of relations associating it with various intracellular functions, so that its presence permits these functions to be reproduced. The first molecules of nucleic acid to have been formed on Earth were not genetic. One must bear in mind that a capacity for self-replication is not, in and of itself, sufficient to make a nucleic acid a piece of genetic material—that is, to make it capable, in one way or another, of producing the properties of the organism that contains it.

The appearance of life, then, is not the same thing as the appearance of genetic information. For in an RNA world, RNA is not genetic and informational in the same sense that DNA is genetic and informational in the present living world. The hypothesis of an RNA world was therefore a crucial step toward abandoning the informational picture of life. New developments in molecular research itself also showed the limits of this picture, which had nourished hopes that the decoding of the genome would give virtually direct access to the secret of life.[7] These hopes have faded somewhat in recent years with the realization that being able to read the book of life has not led to any great or immediate revelation.[8] Nor did the targeted modification of genes always produce the expected results: a gene that was thought to be essential could sometimes be inactivated without the organism showing any ill effects, while a gene that had been thought to be well understood often turned out to have new and unsuspected functions in the organism.[9]

From this point of view, the fate of the idea of a genetic "program" is particularly telling. First used in connection with the operon model—the first model of gene regulation, put forward by Monod and Jacob in 1961—the notion of a program was intended to explain the precise temporal and spatial regulation of gene activity observed during embryonic development and cell differentiation.[10] Taken literally by some biologists, who drew an exact parallel between a genetic program and a computer program, the notion was much criticized.[11] For one thing, it reduced the functioning of the organism to that of its genes, while ignoring both the environment and the structure and content of the organism's cells. Moreover, the analogy with a computer program was flawed, for in the imagined genetic machine of life it is impossible to distinguish between hardware and software (printed circuits and programs), just as it is impossible to distinguish between programs and data. Nevertheless, the term "genetic program" is still widely used in a metaphorical sense to designate genes and the regulatory sequences that control the functioning and development of the organism.

Although the genomic sequences of a growing number of organisms are being worked out more quickly with every passing year, we do not know much more about life than we used to. It has not been possible to automatically put the accumulated mass of information to practical use—that is, to understand and manipulate the observed phenomena. Paradoxically, then, an unprecedented addition to our supply of information about the world has enlarged the potential scope of human knowledge while, at the same time, revealing the depths of our ignorance. Biologists have therefore been forced to abandon the idea that the complex structures and functions of an organism can be known once its genomic sequence has been deciphered. Most researchers today accept that such structures and functions emerge from the coordinated interaction of thousands of elementary components. Whereas the early molecular biologists supposed that order was to be found in the construction and arrangement of the organism's molecular components, scientists now look instead to the self-organization and integrated functioning of these components.[12] Accordingly, the task facing the new post-genomic technologies is to describe organic complexity. Whatever it may be that defines life, that explains life, is in any case not a mere genetic text. The explanation must be sought elsewhere.

This challenge to molecular reductionism comes at an opportune moment. The "reemergence of 'emergence'" is a general cultural phenomenon of our time, extending beyond science and philosophy into the discourse of public health and politics.[13] Indeed, a better example of an emergent property than life itself—something incomparably new and different in relation to all that came before it—can hardly be imagined. The French philosopher of science Anne Fagot-Largeault has shown that emergent phenomena were initially ignored or denied by scientists, and only gradually recognized as facts before finally being transformed into a wide variety of experimental approaches.[14] The attempts of condensed-matter physicists to link microscopic events to macroscopic changes are an example of the kind of work that has recently given the notion of emergence a certain measure of scientific respectability. Biologists (with the honorable exception of researchers in ecology and evolutionary genetics) were slow to acknowledge this paradigm shift, perhaps because for many of them the success of the molecular approach had obscured the need to take emergent phenomena seriously.

I shall return later to the question of complexity and emergence in connection with life. For the moment it is enough to say that these notions have helped to devalue the informational model that previously dominated research in the life sciences. The reassessment of older ideas is probably also due to the fact that a great deal of biology is no longer done by biologists. With the rise of genomics and post-genomics, bioinformaticists, physicists, and mathematicians turned their attention to greener pastures. Not only were these "new biologists" more receptive to the notion of emergence, they had no prior commitment to the informational picture of molecular biology. By itself, however, a shift away from this style of thinking could not have restored the question of life to its former place of honor. The search for extraterrestrial life and the development of exobiology (now called astrobiology) were also necessary.

The idea that life forms might exist on other planets in our solar system, or beyond, was hardly novel, of course. What was new was that with the advent of human space exploration in 1957 it had become possible, or so scientists believed, to tell whether such forms in fact exist.[15] The term "exobiology" was coined almost fifty years ago by the late American geneticist Joshua Lederberg, one of the major figures in the molecular revolution that was then reshaping the life sciences.[16] The principles of life on Earth having apparently been discovered, it seemed natural to ask whether they operated in other environments. In 1976, the first devices specifically designed to look for extraterrestrial life reached Mars as part of the Viking space probes launched by the National Aeronautics and Space Administration (NASA).

The current enthusiasm for the search for life elsewhere echoes the popular fascination with these early projects.[17] Part of the motivation for renewed exploration, which NASA saw as a way of rekindling the American public's fading interest in space travel, was political: the promise of discovering new life forms was just what was needed to put an end to years of congressionally mandated funding cuts. Although the sums involved and the means mobilized for astrobiological purposes are rather small compared with other space projects, they far surpass the level of funding that biologists can normally expect to receive. Their interest was further heightened when NASA announced that particular importance would be attached to the characterization of "extremophiles"—terrestrial life forms inhabiting environments that until recently were considered hostile to all forms of life, but that may in fact resemble ones existing on or under the surface of planets or moons thought to be capable of supporting life. The European Space Agency followed suit and launched its own astrobiology program, Aurore.[18]

Major decisions in science policy must be based on a shared perception of the priority assigned to a given project. Governments have always had to set ambitious objectives (typically described as distant frontiers to be crossed, and so on) to attract broad support from the scientific community no less than from the public. In the 1960s, there was general agreement about the importance of exploring space and, more specifically, manned exploration of the moon. Success in today's mission—investigating the existence of life forms elsewhere in the solar system, and even beyond—would clearly have far-reaching consequences for mankind's conception of its place in the universe.

Further justification for the current program can be found in the experimental evidence that has accumulated since the early 1990s regarding the existence of planets orbiting stars other than the sun. For the moment, given the limits of current technology, it is easier to detect giant gas planets like Jupiter and Saturn, which are not thought likely to sustain life, than other kinds.[19] But research in this direction has just begun, and astronomers are hopeful that it will soon be possible to detect and characterize small, rocky planets the size of our own traveling around the stars nearest our system.[20] Indeed, the first such planet may already have been found.[21]

It is important to note that progress in astrobiological research does not necessarily require answering the question of what constitutes life. Many projects in this field aim only to identify organic samples that have the same structural and functional characteristics as ones currently found on Earth. This is sometimes a deliberate choice, especially where it is assumed that the range of constraints to which all living forms are subject implies that any extraterrestrial organism must exhibit characteristics very similar to those observed on Earth. More often, however, it reveals a certain intellectual laziness, perhaps even a fear of getting lost in philosophical speculation. And yet although many scientists consider inquiry into the nature of life to be a metaphysical rather than a properly scientific enterprise, it seems unlikely that effective research in this domain can be carried out without seriously posing the question—without trying to distinguish between fundamental characteristics that are shared by all life forms, no matter where and when they exist, and accidental characteristics associated with a particular environment.

Part II
The Question in Historical Perspective

My purpose in this second part of the book is not to make a list of all the answers, past and present, that have been given to the question "What is life?" An entire book of such answers would require knowledge that I do not have, and in any case it would not be long enough. I mean instead to bring out the various points of agreement and contradiction hidden among the many different answers that have been proposed.

The American historian of science Gerald Holton suggested that the evolution of scientific ideas could be described as a series of oscillations between pairs of opposed terms, which he calls "themata."[1] In much the same way, Georges Canguilhem spoke of biological thought as having been torn throughout its history between opposing poles, continually moving back and forth between them: "Mechanistic thought and vitalism confront each other over the problem of structure and function; continuity and discontinuity over the problem of the succession of forms of life; preformation and epigenesis over the problem of development of the organism; atomicity and totality over the problem of individuality."[2] It will be useful, then, to try to illuminate these different ways of thinking, which any discussion of the nature of life must take as its point of departure. After all, we cannot think outside the box unless first we have thoroughly inspected the inside.

"Philosophy," according to the French philosopher Maurice Merleau-Ponty, "is the totality of questions in which the questioner is called into question by the question."[3] If so, the question "What is life?" is very much a philosophical question. But it is also a part of science, and sooner or later the answer will come from advances in biological knowledge. Just the same, life is not something external to the world, something tacked onto it—an inanimate material domain that alone can be studied scientifically—and the insistence of some philosophers that life lies beyond the reach of science should not lead scientists in general, and biologists in particular, to suppose that the question "What is life?" is meaningless. Nor should they pretend that any of the answers given so far is truly satisfying.

4

A Rich Heritage

The first answer to the question "What is life?"—or at least one of the oldest and best known—comes from Aristotle: "By life we mean self-nutrition and growth (with its correlative decay)."[1] On this view, life is a temporal process with a beginning and an end.

In *Physiological Researches on Life and Death* (1800), the French physiologist and pioneering histologist Xavier Bichat uttered his famous aphorism, "Life is the totality of functions that resist death."[2] In saying this Bichat affirmed Aristotle's opinion, but he focused less on the question of life's beginning than on its end, describing survival as a struggle against the world and the threats to life that the physical environment contains. This clear description of life in terms of its opposite, death, has its origin in the vitalist tradition that developed among naturalists (and, in particular, physicians) during the eighteenth century in reaction to the mechanistic influence of Cartesianism and its view of bodies as machines. For Bichat, as for many other biologists of his time, each organism contained a principle of life, analogous to Newtonian gravitational attraction.[3] None of them attempted to analyze or describe it, however.

The vitalist conception of life gradually became identified with the chemical and thermodynamical approach that grew up in the nineteenth century and flourished in the twentieth century. In this perspective, life is growth and development, both at the individual level and with regard to the population of organisms as whole. Growth was seen as a result of the capacity of organisms to bring about chemical reactions at moderate temperatures—reactions that chemists can induce only under extreme conditions, and then only very inefficiently for the most part. This capacity was thought in turn to be due to the existence of specific catalysts known as diastases (subsequently called enzymes and then, in the mid-1920s, identified as proteins). The "enzymatic" theory, dominant in the early decades of the twentieth century, before the molecular biology revolution, held that the distinctive characteristics of living organisms were associated with the action of these extraordinary catalysts.[4]

According to the second law of thermodynamics, however, the universe as a whole tends toward disorder, or more exactly, toward the uniformization of its properties: entropy, in other words, increases. Organisms appear to violate this law, however, exhibiting what Schrödinger rather unfortunately called "negative entropy." In the second half of the twentieth century, research on the thermodynamics of nonequilibrium states made it possible to resolve this paradox. Although organisms are in thermodynamic disequilibrium, because they continually exchange matter and energy with the external environment, the universe as a whole, including both organisms and their environment, becomes increasingly disordered over time.

All that remains today of the ancient debates over what came to be known as the thermodynamic paradox of organisms is the idea that life is exchange. The ability to exchange matter and energy with the external environment is now generally accepted to be one of life's defining characteristics. When the Greek philosopher Heraclitus likened an organism to a flame, he meant to call attention to the fragility of life. But this analogy has now acquired another meaning: like an organism, a flame needs a continual supply of matter and energy. A flame, like an organism, is an open, self-sustaining system.

A second group of answers to the question "What is life?" which also traces its origins back to Aristotle, is associated with the notion of an organism—the very term he used to describe living beings. Aristotle taught that living beings are characterized by the complexity of their structural and functional organization, which is to say by the existence of different components that act in concert. Developed by Immanuel Kant, Jean-Baptiste Lamarck, and the French naturalist Georges Cuvier in the late eighteenth and early nineteenth centuries, this conception subsequently appeared in the twentieth century as a thread running through the writings of many biologists, whose research yielded an increasingly rich picture of the organizational complexity of living matter. Protoplasm and colloids, two notions that in the second half of the nineteenth century and the beginning of the twentieth century were used to describe the peculiar organization of cellular matter that was thought to account for the extraordinary properties of life, are distant relatives of this conception. The idea of an aperiodic crystal, advanced by Schrödinger in *What Is Life?* to explain the origin of the order displayed by the living world, is also a direct descendant, albeit one that has been transformed by the modern view that genes contain the governing principle of life.[5]

The final point in the development of this idea was reached with the notion of genetic information. It now became possible to overcome a difficulty that had already been emphasized by the French physiologist Claude Bernard in the nineteenth century, and that the early geneticists had simply ignored because they had no answer for it, namely, how is it that order appears during development? If order is already contained in the egg, then what form does it assume there? As François Jacob was later to point out, the idea of a genetic program allowed two equally unsatisfactory solutions to be reconciled: epigenesis (the spontaneous appearance of complexity during development) and preformationism (the integral preexistence of complexity).[6]

Within this second tradition, however, two different conceptions of organic phenomena came into conflict. The first laid emphasis on the material basis of the living world: in order to understand life, it must be possible to describe its component elements. On this view, championed by Bernard, the matter from which organisms are made has "vital" characteristics. The second view, by contrast, laid emphasis on the nature of the relations obtaining between these components: it is the way in which the constitutive matter of organisms is organized that accounts for their vitality. Thus Cuvier argued that because life is characterized by a continual exchange with the environment an organism remains the same organism throughout the course of its existence, notwithstanding that it is never composed of exactly the same material components from one moment to the next. This argument recalls the ancient notion (once again due to Aristotle) of a *form* that transcends the nature of the physical object necessary for its manifestation.[7] The metaphor of Theseus's ship (first mentioned by Plutarch) is still telling: the planks of a boat may be replaced one by one and yet it remains the same boat, even if ultimately the original boat has nothing materially in common with the one that has been rebuilt. This image also tells us that the characteristic of persisting or enduring form is not restricted to organisms.

A third, more recent group of answers to the question "What is life?" deriving from the work of Charles Darwin, holds that organisms are characterized by their ability to reproduce with variation, and therefore to adapt to a changing environment. The idea that living beings reproduce with variation is almost surely correct: it is difficult to imagine any real world in which a process as complex as reproduction could be perfect, and so fail to generate variation. Darwin argued that limited environmental resources led to competition between organisms and to the survival of the fittest ones, which, in turn, transmit to their offspring the characters that enabled them to survive. Population geneticists gradually developed a more nuanced view of natural selection: in order for the frequency of a given form of a character to increase in a population, all that is required is that individuals who possess that form produce a greater number of offspring than individuals who do not possess it. Later I shall return to the Darwinian tradition, which is currently so influential, in order to demonstrate its inadequacy as a basis for a satisfactory definition of life. For the moment, however, we need simply to note that this tradition is related to the two others: by emphasizing reproduction it implies that something—an *organized* something—is transmitted; and because any reproductive process involves growth, and thus positive transfers of matter and energy with the external environment, it also includes the idea of exchange.

But the Darwinian conception also represents a reversal of perspectives and priorities. This can clearly be seen in the history of research on bacterial viruses, or bacteriophages.

Both Félix d'Herelle (the codiscoverer of the bacteriophage) in the 1910s and Max Delbrück in the late 1930s thought that despite its small size the bacteriophage was indeed alive, and therefore a good model for the study of the fundamental characteristics of life. Yet their reasons for thinking this were diametrically opposed. D'Herelle believed that life was characterized by two properties, assimilation and adaptation, which he claimed to be able to demonstrate in the bacteriophage.[8] The first characteristic, assimilation, was nothing other than what Aristotle had called the ability to grow; the second, adaptation, referred to what d'Herelle called the "inventiveness" of life, which he associated with the existence of a vital force. For Delbrück, a Darwinian, the fundamental feature of life possessed by the bacteriophage was the power to reproduce with variation. To his way of thinking, the two aspects of life emphasized by d'Herelle were mere consequences, since reproduction requires assimilation and variation permits adaptation once natural selection has taken place.

The Darwinian conception of life therefore depends upon the existence of autonomous organisms having a line of descent, and denies the possibility of spontaneous generation. (It is interesting to note, by the way, that the debate between Louis Pasteur and the French naturalist Félix-Archimède Pouchet over spontaneous generation took place the same year that Darwin published *The Origin of Species,* in 1859, even though there was no direct link between the two events—indeed, Pasteur was not a Darwinian.) It should also be noted that Darwin's theory grew out of the gradual realization during the early decades of the nineteenth century that all organisms are made up of cells and that, in the German pathologist Rudolf Virchow's famous aphorism, "omnis cellula e cellula" (every cell comes from another cell).

It is clear, then, that these three different conceptions of life—as chemical activity and exchange, as organization, and as the ability to reproduce with variation—are not mutually exclusive, but instead are related to one another in complex ways. There is an obvious objection, of course, to this way of presenting the matter (quite apart from the possibility that one or more of these conceptions have been oversimplified, or even caricatured), namely, that it mixes up science and philosophy. But we need to keep in mind that the divide between science and philosophy was not always so clear as it seems to us today, and also that scientific theories, by their very nature, are never wholly free of philosophical presuppositions or implications. We will see later, for example, that the Darwinian conception of life is often preferred more on philosophical than scientific grounds.

The close relationship between philosophical and scientific approaches has had two consequences: first, that the philosophical heritage of reflection about life is a rich one; second, that when scientists philosophize they often do no more than repeat, generally less effectively and less clearly, what philosophers have already said. Certainly it is true that a fresh view of a problem is often useful, but only so long as one has first taken into account what earlier thinkers have had to say. On the other hand, we should not imagine that everything has already been said. Often when I lecture on the question "What is life?" and the challenge it represents for contemporary biology, no matter whether the audience is made up of scientists or laymen I encounter the same difficulty: many of my listeners adamantly maintain that the question was settled long ago (by Aristotle, Bichat, Bernard, or whomever). But these great thinkers have not succeeded in explaining what life *is;* they show us only what life was once thought to be, for progress in biological knowledge continually changes the answer. Consider, for example, the conception of life as organization. What Lamarck described in a very general and simplistic way is now understood with a degree of detail and precision that Lamarck himself would find absolutely amazing were he alive today. Just so, the vast body of knowledge that supports the concept of organization cannot help but affect the concept itself, and for this reason alter the way we answer the question of what constitutes life.

5

Contemporary Answers

Although most biologists no longer think that looking for an answer to the question "What is life?" is either meaningless or, worse still, irrational, they are apt to be surprised by the difficulties involved. The American biochemist Daniel Koshland described a meeting of eminent figures in the field that was devoted to finding an answer that could be stated in a single sentence. Each time a new definition was proposed, it took the members of the audience only a few minutes to demonstrate its inadequacy. Finally, after several hours of debate, one answer met with general approval and was adopted by almost unanimous consent: the essence of life is the ability to reproduce. A few minutes later, however, some bright spark pointed out that on this view a rabbit is not alive: two rabbits (one male, one female) are alive, but not either one by itself. Coming up with a plausible answer that can be summarized in a few words is not as easy as it may seem.[1]

Any satisfactory reply must also avoid the trap of merely codifying the current state of scientific knowledge, since any such formulation would inevitably run the risk sooner or later of becoming outdated. A historical example will illustrate the problem. In 1894, Friedrich Engels—together with Karl Marx, the father of dialectical materialism—proposed the following answer: "Life is the mode of existence of albuminous bodies, and this mode of existence essentially consists in the constant self-renewal of the chemical constituents of these bodies."[2] Although constant self-renewal does indeed characterize life, the reference to protein structures ("albuminous bodies") renders the definition obsolete. In Engels's time, proteins were one of the few macromolecular organic components that were relatively well understood—hence his emphasis. Today, a similar version of this definition would have to accord nucleic acids at least an equal place.

Nearly a century after Engels, the American molecular biologist Gerald Joyce put forward the following definition, now used by NASA for its exobiology program: "Life is a self-sustaining chemical system capable of undergoing Darwinian evolution."[3] The plausibility of this claim can be evaluated against several possible criteria. First, faced with an unknown object, would it enable us quickly to determine whether or not the object is alive? It is particularly important for the design of missions looking for extraterrestrial life that this condition be satisfied. In other words, the definition must be operational. But "operational" may mean different things to different groups of researchers. A second criterion is the consistency of the definition with what we know of life on Earth: it must not exclude any known life form, while excluding all non-organic objects. Finally, the answer must be internally coherent and succinct, but sufficiently rich to capture much of the specificity of life.

Joyce's definition has the advantage of being simple while nonetheless distinguishing two fundamental characteristics of living organisms: they are chemical systems, and they are capable of undergoing Darwinian evolution. Nonetheless it suffers both from a lack of precision and from certain crucial omissions. What exactly does the phrase "self-sustaining chemical system" mean? Does it refer merely to the temporal persistence of chemical reactions? Or does it refer to the capacity for self-renewal of some or all of the molecules involved? The phrase "capable of undergoing Darwinian evolution" is unsatisfactory as well, insofar as it seems to define an organism more in terms of a potential capacity than an actual property.

Nothing is said about the autonomy of living systems, for example, or about the property of being bounded by a membrane. Even so, the ideas of autonomy and closure are not completely absent: they are hinted at not only by the term "self-sustaining" but also by the reference to Darwinian evolution, which implies the existence of autonomous living organisms capable of producing offspring and of evolving by adaptation. What is more, the vagueness of Joyce's definition gives it the advantage of not being restricted by the limitations of current knowledge. It says nothing, for example, about the content of this self-sustaining chemical system—but if tomorrow we were to discover that the RNA world was preceded by still another world based on a different polymer, this definition would survive intact.

Surprisingly, however, since it has been adopted by NASA, Joyce's formula might not in fact be useful in the search for extraterrestrial life forms. Even if the existence of chemical reactions in a given object could be demonstrated, it would not be easy to prove they were self-sustaining. It is still less clear what kind of experiment would be able to detect the presence of a chemical system capable of undergoing Darwinian evolution. And while this formula includes all known terrestrial life forms, what are we to conclude about viruses? Are they organisms? Viruses are indeed susceptible to Darwinian evolution, but are their chemical systems self-sustaining? The parasitic nature of viruses would appear to disqualify them, but the inexactness of this formula prevents us from drawing any firm conclusion. On the other hand, even though Joyce does not specify any particular kind of chemical reaction (for example, ones involving carbon), he does exclude all forms of life that do not arise from some set of chemical reactions, such as artificial life forms created by a computer.

Despite its weaknesses, Joyce's definition succeeds in establishing that any satisfactory answer must at least take into account two distinct characteristics: the chemical properties of organisms and the role of Darwinian evolution. Another answer has recently been proposed by two young French biologists, Patrice David and Sarah Samadi. Although their main interest is the theory of evolution, they became aware of the need to come to terms with the problem of what constitutes life because, in order to construct such a theory, one "must first set limits to its field of application, by providing a definition that makes it possible to distinguish organisms from other objects."[4] An admirable purpose—if only because most biologists do not even bother to try! It turns out, however, that David and Samadi have a vested interest when it comes to analyzing the properties of life: for them, only the theory of evolution provides a basis for investigation.

Rather than follow Joyce in giving a brief and pithy answer, David and Samadi proceed by first identifying the properties that characterize life and then deciding which of them is essential.[5] Living organisms, they say, share three crucial properties: a distinctive molecular constitution that differs from that of other physical objects; a complex architecture that is maintained despite the constant transformations of the material substrate; and the capacity to reproduce, due to the presence of informational molecules (in particular, nucleic acids). The next step is to decide which of these properties constitutes a universal criterion for recognizing organisms in an environment different from that of Earth. David and Samadi take the view that reproduction alone meets this standard. The other characteristics—products of the particular history of the environment in which organisms developed on our planet—they regard as only contingently associated with life.

This is an interesting way of approaching the question. One begins by deciding what will constitute a satisfactory answer. Any such answer, these authors hold, must be universal, applicable to every possible set of local conditions in which life may appear. But rather than give an answer a priori, they prefer instead to draw up an ordered list of characteristics. This, by the way, is the very same approach recommended by Claude Bernard in *Phenomena of Life Common to Animals and to Plants* (1878): decide first on the characteristics of organisms, then put them in their natural hierarchical order.[6] In doing this, the focus is shifted from a purely descriptive definition to a causal explanation of the unique characteristics of organisms. The characteristics David and Samadi have chosen are by no means original; indeed, the second characteristic (organization that survives the transformations of matter) would not be out of place in the writings of Cuvier. There is nothing wrong with this, of course, but the reasons for giving pride of place to reproduction as the universal property of life are not fully explained. It is perfectly legitimate to argue that the matter from which organisms are made depends on the environment in which they are formed, and to conclude from this that the first characteristic is less fundamental than the others. But it is more difficult to see how the environment might be responsible for the capacity of organisms to evolve a complex, stable architecture. On balance, the claim that reproduction is the key determinant of life must be considered unpersuasive, for reasons that I shall examine more closely later.

David and Samadi are also on weak ground in directly tying the power of reproduction to the presence of informational molecules, in particular nucleic acids. As we have seen, the term "informational" is extremely ambiguous. Here again a problem arises in trying to understand the precise nature of the link being asserted between reproduction and the existence of informational macromolecules. It would appear that the existence of such molecules is assumed to be a necessary condition of reproduction. If so, how are informational macromolecules replicated during this process? Did molecular replication precede reproduction, or did the two appear simultaneously?

The list of distinguishing properties given by David and Samadi does not include a number of characteristics that have traditionally been associated with life and with organisms. These characteristics are explicitly mentioned, however, by another pair of authors, the Spanish philosopher of science Álvaro Moreno Bergareche and his compatriot Julio Fernández Ostolaza, an artificial life researcher.[7] Most specialists in artificial life were originally trained in artificial intelligence, which uses computers to reproduce behaviors that were previously considered unique to humans.[8] These scientists are therefore well equipped to make a major contribution to the subject at hand, for the simple reason that the first step toward creating artifacts that mimic the vital characteristics of organisms consists in defining the characteristics that one wants to imitate. Accordingly, though Fernández and Moreno do not directly answer the question "What is life?" they do provide a list of properties that are intimately associated with it. This is a much longer list than the one given by David and Samadi, and the properties it mentions are not put into any kind of order:

- Life implies the existence of a spatio-temporal structure. (A vague formulation, it is true, but the authors go on to say that in a complex organism a general renewal of its cells takes place that preserves the organism's overall form. This is the ancient Greek metaphor of the boat, only now explicitly applied to the global, macroscopic organization of the organism.)
- Life requires processes of reproduction.
- Organisms store information that encodes a description of themselves.
- All organisms possess a metabolism—a set of chemical reactions—that converts matter and energy from the environment into elements and energy that suit the needs of the organism.
- Complex organisms functionally interact with their environment.
- All organisms exhibit interdependence among their components.
- All organisms attempt to stabilize themselves in the face of perturbations.
- Life evolves.

While acknowledging that their list is far from complete, Fernández and Moreno point out that any further properties which might be added would necessarily be related to two properties that they see as essential to the existence of life: autonomy and the ability to process information.

This account differs in certain respects from the two we have already looked at, those due to Joyce and to David and Samadi, neither of which mentions the characteristic of resistance to external variation, for instance. Furthermore, the relation between the list of properties as a whole and the two properties that Fernández and Moreno take to be fundamental is unclear. In particular, they do not say whether organisms can be defined simply as autonomous units capable of processing information, with all the other properties being implied by this definition. If so, their account would depart from the other two in not recognizing the ability to reproduce as one of the fundamental characteristics of life. What distinguishes their non-exhaustive list of properties from the others is that it is based on the study of "higher" organisms and their properties. There is nothing surprising about this: from its inception, as I say, research on artificial life has been closely associated with work on artificial intelligence. Von Neumann himself, in the years before his death, was interested in the possibility of creating both electronic calculating machines that could think and self-reproducing automata.

There is a philosophical justification for this long list of properties associated with life as well. For Fernández and Moreno, the fact that some properties are found only in the most complex organisms does not mean that they should be ignored; to the contrary, their very existence shows what life may lead to—what its potential was from the very beginning. This was Claude Bernard's position as well: properties that appear only at the apex of evolution are just as fundamental as those that are present in all organisms. The rich diversity of organic forms and structures observable in nature (a criterion that is not mentioned here, by the way, but which figures in comparable answers) can also be regarded as a fundamental property of life, notwithstanding that each organism is only one of the many components that produce this diversity, which gradually increases during evolution. I shall return later to this philosophical argument in favor of taking into account more recent products of evolution.

Finally, let us consider a fourth answer, proposed by the British biochemist J. Perrett in 1952: "Life is a potentially self-perpetuating open system of linked organic reactions, catalyzed stepwise and almost isothermally by complex and specific organic catalysts which are themselves produced by the system."[9] This answer corresponds to half of the answer given by Joyce, but it is stated more precisely. Moreover, it is representative of a class of definitions among which might also be included the "autopoietic" models of self-organization developed by the Chilean neurobiologists and philosophers of science Humberto Maturana and Francisco Varela.[10] Perrett's answer has the virtue of being less abstract than these models, which hold that the special character of life can be found in the nature of the dynamic relations that link the components of living systems, the precise chemical properties of which are not considered to be important. What is particularly noteworthy here is the complete absence of any reference in Perrett's definition to reproduction: for him, as for Maturana and Varela, the ability of organisms to reproduce is secondary—in which case a single rabbit is indeed alive!

The American biophysicist Harold Morowitz has argued that this type of answer—the earliest of the four contemporary definitions we have looked at, at least in its original formulation—remains the most satisfactory one.[11] Still, we cannot lightly dismiss the idea that the primary property of organisms is their ability to reproduce (with variation), for it occupies an important place in the thinking of present-day biologists and, historically, has deep roots in Darwinian theory.

These four answers are not the only ones that have been given. Nor are they necessarily the best ones, or even the most recent.[12] I have chosen them because they very nicely illustrate the difficulties that are repeatedly encountered in this connection: Must a satisfactory answer include a description of the components of organisms? Is the ability to reproduce fundamental? Can we speak of life in the absence of autonomous organisms? Can an adequate description of life be based only upon the simplest organisms? The answers to these questions are interrelated as well. If the ability to reproduce is considered to be primary, for example, then the nature of the components that form organisms becomes secondary. Additionally, the ability to reproduce is associated with the acquisition of autonomy, and it is already present in the simplest life forms.

The sort of answer one gives to the question "What is life?" is apt to depend on one's professional training as well. Chemists and biochemists may naturally be expected to assign greater importance to the metabolic aspects of life, molecular biologists to the role of informational macromolecules, and population geneticists to the ability of organisms to reproduce and evolve. This disciplinary bias is hardly surprising: a scientist whose entire career has been spent studying metabolic reactions is very likely to consider them fundamental. Just the same, the influence of disciplinary orientation should not be exaggerated. Scientists' views, particularly with regard to the question of what constitutes life, tend to reflect fundamental differences of opinion rather than allegiance to a particular school of thought or line of research. And with the blurring of boundaries between the different branches of biology in recent years, it is often difficult for a biologist to know whether she or he should be called a biochemist, a molecular biologist, or a geneticist.

The various answers we have been considering nonetheless justify us in concluding that organisms share three essential characteristics: they can reproduce; they possess complex molecular structures; and they exhibit intense metabolic activity that leads to the replication of such molecular structures. These three characteristics coincide to some degree with the three conceptions of life that emerged from our historical study in the previous chapter, and will help us state them more precisely. Most biologists would probably agree that these three characteristics are extremely important. This agreement would vanish at once, however, if they were asked to rank them in order of priority, or to eliminate one or two of them, or—what amounts to the same thing—to answer any of the questions that I raised a moment ago with regard to the conditions an adequate answer must satisfy.

Part III
Contributions of Current Research

In this part I take up answers to the question "What is life?" that grow out of present-day biological research. To some extent these are what might be called subliminal answers, in the sense that scientists do not feel obliged to discuss them openly, generally because they are taken for granted. Nonetheless scientific work would lose its meaning without them.

There are some cases where the answer, even when it is not stated explicitly, is plain enough. For example, the experimental apparatus placed on a space probe to detect traces of life must have been conceived and developed with a clear idea of what life is, and so of what its distinguishing characteristics are. Research aimed at reconstituting the chemical and biochemical stages by which life appeared on Earth, in order then to construct a realistic scenario of its origin, is likewise based on a particular conception of life. In both cases scientists have selected a number of characteristic features, from the many that are imaginable, that they think must have constituted the rate-limiting step in the formation of life. The notion of a rate-limiting step comes from studies of chemical reactions in metabolic pathways: one reaction always occurs more slowly than the others and, by forcing the others to obey its rhythm, puts a brake on the metabolic activity of the system as a whole. Generally speaking, this step is also the one that is the most finely modulated, and therefore the most important. Studies of the origin of life owe part of their interest, it should be noted, to the ambiguity—conscious or unconscious—of the word "origin," which signifies both a beginning and a foundation; birth, but also the conditions responsible for it.

All biological research, even if it is not directly concerned with the origin of life on Earth or the search for extraterrestrial life, is implicitly based on a particular conception of life. Even if only very indirectly, and in no more than a fragmentary manner, its purpose is to reveal the basis of organic phenomena—to explain life and the order that can be observed in the living world. This is especially true of studies that seek to determine, or to extend, the ecological limits within which present-day organisms have developed and the genetic constraints to which they are subject. I shall begin by discussing research aimed at reconstituting the path of development followed by life on Earth, and then look at research on life as we know it, including life under extreme conditions. In the penultimate chapter of this section I consider the rapidly expanding field of astrobiology.

6

Looking for Life's Past

Trying to reconstruct the path that led to the appearance of life on Earth makes sense only if terrestrial life first appeared—on Earth. There is a competing hypothesis, known as panspermia, put forward first by the Scottish physicist Lord Kelvin in the latter part of the nineteenth century, and subsequently developed by the Swedish chemist Svante Arrhenius in the early twentieth century.[1] According to this theory, life has always existed in the universe and has simply moved from planet to planet. The theory appeared to have been definitively refuted in 1924 by the French plant physiologist Paul Becquerel, whose experiments suggested that no organism could survive the low temperatures, the vacuum, and above all the ultraviolet radiation that exist in interplanetary space.[2]

Until recently, the panspermia hypothesis seemed to have all but disappeared. It has been revived for a number of reasons, initially owing to the discovery more than forty years ago that various molecules out of which organisms could be built (ammonia, cyanhydric acid, formaldehyde, and even relatively simple biological compounds such as amino acids and sugars) are present in surprisingly high concentrations in the universe, particularly in the meteorites and comets that regularly strike Earth.[3] By itself this observation could not prove the hypothesis, but it did show that the basic elements of life are found throughout the universe. This was not actually news. In May 1864 a meteorite hit France and, despite the limited analytical methods available at the time, chemists were able to show that it contained organic matter, probably of extraterrestrial origin. Pasteur himself examined the debris but could find no trace of microorganisms.[4] More recently, Francis Crick, the codiscoverer of DNA, lent his support to a version of the theory of panspermia, which in part explains its resurgence.[5] In the 1990s, the Cambridge astronomers Fred Hoyle and Chandra Wickramasinghe argued in a series of books that life came to Earth from outside the solar system and that extraterrestrial life forms are continually entering the biosphere.[6] The announcement at about the same time that apparent traces of life had been discovered in a meteorite from Mars seemed to confirm the possibility that life arrived on Earth from elsewhere.[7]

For the moment, however, interest in panspermia seems once more to have receded. Arguments adduced in support of the claim that traces of life were present on the Martian meteorite—a discovery, trumpeted to the media worldwide by NASA, that came just as the U.S. space agency was beginning to allocate a large part of its budget to astrobiology programs—were disproved in rapid succession.[8] The panspermia hypothesis was more plausible on the assumption of an eternal (or steady-state) universe like the one that was generally thought to exist by astronomers in the early twentieth century, by contrast with the present-day view of a universe in constant expansion following an initial explosion (the Big Bang). It is now accepted that the universe has a definite age, which makes it unlikely that the formation of other planets preceded the Earth's appearance by a long enough time to be of any help in explaining the origin of life. Indeed, the suggestion that life evolved on a faraway planet of which we know nothing has the grave disadvantage of making the origin of life unknowable, at least for the foreseeable future.

If we assume that terrestrial life did in fact originate on Earth, we need then to try to reconstruct the path leading to its formation and development. There are three possible ways of approaching this problem: first, searching for early traces of life on the surface of the Earth; second, breaking down the process by which life was formed into a series of steps and then showing how each one came about; third, examining present-day organisms for traces of previous states that represent the earliest forms of life, or possibly even prebiotic states.[9] The second and third of these approaches I will take up in following chapters.

The first approach amounts to applying the classical methods of paleontology, which seeks to characterize the fossil remains of extinct animals or plants, to fossil microorganisms. The age of these earliest forms of life on Earth makes this particularly difficult, and requires micropaleontologists to extrapolate very carefully from what is known about the age of the various rocks that litter the Earth's surface. The presence of mineral concretions that are formed only by the presence of living organisms, such as stromatolites, may also be a sign of primitive life forms. It is possible, too, that traces may be found, not of individual organisms, but of colonies of bacterial microorganisms whose distinctive forms would provide evidence of specific reproductive processes.

Increasingly sensitive physicochemical techniques hold out the prospect of one day being able to detect the presence of chemical signs of life associated with these physical traces: either molecules specifically produced by life (or by a particular form of life), or a slight imbalance created by certain chemical reactions set in motion by life (between the various isotopic forms of an element, for instance, given that organic molecules incorporate a particular isotope of carbon).

This kind of research involves many difficulties and potential sources of error. Very ancient geological strata are hard to identify, and may be contaminated by more recent organic material. The older the rocks, the greater the risk that they have been altered in the vast interval between past and present. It is also very difficult to distinguish visually between the genuine trace of a bacterial colony and what is only a mineral artifact. Complicating matters further, natural chemical processes in inorganic material may lead to the production of particular molecules or isotopic imbalances that until now have been considered signs of life.

Recently there has been a series of spectacular announcements, each one pushing the origin of various life forms on Earth further back in time: 3.85 billion years for the first traces of bacteria, 3.7 billion years for the earliest evidence of photosynthesis, 2.7 billion years for the first remains of complex eukaryotic cells (cells containing a nucleus), 1.2 billion years for the first multicellular organisms, which were kinds of worms.[10] But virtually all of these results have since been challenged for any one of a number of reasons—errors in geological description, confusion of artifacts produced by crystallization with forms specific to life, the discovery of chemical mechanisms capable of producing isotopic imbalances in the absence of any life form, and so on—and the established chronology has not been fundamentally altered.[11] New observations have confirmed that the first life forms appeared at least 3.5 billion or so years ago, and the first eukaryotic organisms around 2 billion years ago.[12] The first multicellular organisms are still thought to be only slightly more than 1 billion years old.

Researchers are hoping to find traces of cellular life encapsulated in a membrane, which is to say a microorganism similar in size and structure to microorganisms found on Earth today. It is reasonable to suppose that the earliest forms of life were very small, for in that case there would have been few impediments to the spontaneous movement of molecules through diffusion. Furthermore, the precise division into two virtually identical daughter cells is accomplished more easily in a small structure than a large one. The search for a cellular form of life proceeds from a recognition not only of the importance of isolation and self-organization in the formation of the organic world, but also, and especially, of experimental necessity: identifying the traces of something that resembles a cell is fairly simple, whereas proving the existence of noncellular life, even with the benefit of the most sophisticated physicochemical techniques, is virtually impossible, for the danger that later life forms had contaminated the sample can never be completely excluded.

Following its formation around 4.5 billion years ago, the Earth was continually bombarded by meteorites for several hundred million years. A terrestrial environment capable of supporting life could scarcely have come into existence earlier than four billion years ago. Had life appeared before this time, the subsequent modifications of the Earth's surface would surely have erased all traces of it unless initially it had developed deep below the surface. In geologic time, then, the earliest organisms probably appeared shortly after the Earth's formation. If we accept the hypothesis that the living world we know today was preceded by other living worlds, in particular by an RNA world, the reign of these first life forms must have been very brief.[13] Complex cells did not appear until more than another billion years had passed, and multicellular organisms not until another one billion years after that.

Many biologists and astrobiologists regard this chronology as evidence that the formation of simple organisms is a relatively straightforward process, and that such cellular forms of life are therefore likely to be widespread throughout the universe. The formation of cells having more complex structures and, above all, the formation of multicellular organisms are commonly thought to be more problematic, however, and thus quite unlikely to be widespread.[14] As for the chance of discovering complex life forms on other planets, ones with cognitive abilities equal to or greater than those of human beings, this is thought to be slim, if not effectively zero—a very different view from the one expressed by the myth of Frankenstein's monster: once Frankenstein had discovered the secret of life, he encountered few difficulties in creating a living human creature. Two centuries ago—for Mary Shelley, at least—the main barrier to be crossed lay between life and non-life.[15]

My own feeling is that such conclusions may be somewhat premature. First, they are based on dates that, as we have seen, are uncertain. Second, the time between the moment when an event becomes possible and the moment when it actually occurs is converted by the conventional wisdom into a measure of the event's probability. This kind of calculation is virtually worthless, considering that we know so little about the mechanisms that led to the formation of life on Earth. If the formation of unicellular or multicellular organisms are random processes, the sole instance of the appearance of life to which we can point—its appearance on Earth—tells us nothing about the probabilities associated with other possible instances. On the other hand, if these processes are slow but nonetheless highly probable, so long as there is enough time for them to occur, still we can deduce nothing about the probability of the formation of life—unicellular or multicellular—elsewhere in the universe on the basis of terrestrial observations, for the simple reason that we do not really know anything about the conditions under which these processes occur, nor about the probability that these conditions exist on other planets. Furthermore, the move from chronology to probabilities is based on the assumption that life (whether microbial or something more complex) has appeared only once. But it may be that life has appeared several times, vanishing only to reappear later in a new form.

Given such uncertainties, it is perhaps surprising that the current chronology should nevertheless have been overinterpreted and, in particular, such strong conclusions drawn with regard to the nature of life forms that may (or may not) be found on other planets. It is true that a number of recent observations lead to similar conclusions, at least indirectly. First, a hitherto unsuspected wealth of diversity has been discovered among microorganisms. Only thirty or so years ago, Carl Woese shook the foundations of microbiology by showing that the living world is divided into three branches, rather than two as previously thought: eukaryotic (nucleus-bearing) organisms, whether unicellular or multicellular; traditional bacteria, to which Woese gave the name "eubacteria"; and a whole new set of bacteria called "archaebacteria" (now known as "archaea"), as different from traditional bacteria as they are from eukaryotes. Moreover, discoveries in the past few years have demonstrated that the diversity of bacterial life—eubacteria and archaebacteria—is far greater than previously imagined.[16] From the earliest days of microbiology, microorganisms had to be cultured in order to produce sufficient quantities for purposes of analysis. But with the development of the polymerase chain reaction (PCR), a technique for amplifying DNA, it became possible to characterize bacteria individually and to compare them directly to known bacteria. As a result, researchers realized that many very small microorganisms with poorly defined shapes were unable to grow on any of the culture media used, thus escaping detection. These microorganisms can live in extreme conditions, miles beneath the surface of the Earth or of its oceans, with a very low metabolic activity.[17] The "Woesian revolution" not only expanded the richness of the microbial world; it also upset the traditional hierarchical view of organisms, according to which their complexity is linked to their order of appearance during evolution. The evolutionary record of life is neither simple nor well traced: there is no path leading from the simple bacterial cell to the more complex eukaryotic cell; instead there is a complicated—and, as we shall see later, not yet complete—scenario in which present-day bacterial cells are no less fully developed than eukaryotic cells.

Second, the world of complex organisms—in particular, mammals—has turned out to be more sensitive to variations in the terrestrial environment than had previously been supposed. A period of explosive growth in the development of complex animal forms, including the development of the main body-plans of current and extinct species, took place in the space of a few million years at the beginning of the Cambrian period, around 530 million years ago, probably under the joint effects of an increase in the level of oxygen in the atmosphere, a change in sea level, and perhaps the disappearance of earlier life forms following a series of devastating meteorite strikes.[18] Many forms of life that appeared during the Cambrian disappeared shortly afterward. Similarly, multicellular organisms well adapted to terrestrial conditions, such as the dinosaurs, vanished as a consequence of the harsh climatic changes produced by meteorite impacts.[19] The appearance of mammals was highly dependent on changes in the environment, and the risk of their sudden extinction was high as well.[20] This sensitivity of complex multicellular organisms to environmental conditions, particularly to the amount of oxygen in the atmosphere (itself the result of the activity of microorganisms that, in splitting water molecules, free oxygen atoms), suggests that the emergence of such organisms was a relatively unlikely phenomenon, due to the chance occurrence of external events.

Finally, there is the argument known as Fermi's paradox, after the famous Italian-born nuclear physicist who stated it in 1950. If the evolution of life leads in a relatively uncomplicated way to the development of complex organisms, with an intelligence equal or superior to our own, extraterrestrial civilizations created by these beings should have developed the means to travel between the stars. But in that case, as Enrico Fermi is said to have asked in conversation one day, "Where are they?" Why is it that in the whole of human history there is no credible trace of contact with an alien civilization? The psychological and sociological explanations put forward to explain the aliens' silence (lack of curiosity, a desire to remain undetected, and so on) seem rather implausible. It may well be, then, that extraterrestrial civilizations are exceedingly few in number, or even nonexistent, and that the evolution of life toward complexity and thought is an event of extremely low probability.[21]

Yet these arguments do not necessarily justify us in supposing that, whereas bacterial life is widespread throughout the universe, complex life forms are rare. Scientific facts can always be interpreted in a number of ways. It can be argued that the Cambrian explosion, like the evolution of life forms in the wake of mass extinctions (such as the rapid expansion of mammals following the disappearance of the dinosaurs), shows the inventive potential of evolution and points to the probability, rather than the improbability, of the appearance of complex life forms. We know, for example, that the phenomenon of cephalization and the evolution of intelligent forms of life did not occur only in the human branch of the terrestrial animal kingdom.

The conviction that only simple life forms are widespread in the universe has metaphysical implications as well. For if we consider that, although the formation of bacterial life is quite probable, the appearance of complex life is relatively unlikely, we introduce a rupture in the continuum of life that has long been supposed to unite the simplest and the most complex life forms. Such a view would run counter to a whole tradition of philosophical and religious thought, which holds not only that life must lead to complexity but that the end point of this development—the emergence of self-consciousness—is contained in the germs of the first life forms, and even in inanimate matter. This tradition remains dominant in continental philosophy, particularly in phenomenology.

Perhaps it would be both more reasonable, then, and more scientific as well, in the true sense of this word, to wait a while yet before we pronounce on the question of whether the appearance of life forms, be they simple or complex, is either very probable or very unlikely.

7
Retracing the Path of Life

The search for traces of the earliest forms of life on the Earth's surface cannot hope to go beyond these forms, however, all the way back to the period of prebiotic evolution that preceded them. The attempt to do just this—the second of the three possible approaches for reconstructing how life was formed—has given rise to a great many investigations since Stanley Miller's 1953 experiment demonstrating the formation of amino acids from simple molecules.

These studies have had two different aims, which are not always clearly distinguished. The first is to describe the path followed by life during its formative phase. This is very difficult, given the paucity of information that is, or is likely ever to be, available about what happened. Indeed, it is probable that there was no single linear trajectory of the sort we associate with a path in the usual sense. (There is not the slightest chance that a memorial plaque will one day be set up somewhere on Earth bearing the inscription: "Here the first autonomous living cell appeared, *x* million years ago.") Most studies have therefore had another, quite separate purpose. They are intended to establish, as precisely as possible, a series of events that might have led to the formation of life, and then to verify that the steps posited by this scenario agree with physiochemical laws and what we know of terrestrial life. It is not unreasonable to imagine that some of these steps could be carried out, at least in part, in the laboratory. Whatever the correct scenario turns out to be, it is probable that some of the events it describes are still occurring on Earth today, and that the omnipresence of life immediately interferes with their further development.

The simplest way of presenting these studies is to start from what might be called the standard account of the origin of life, which has been considerably refined since Miller's original work. In this account, the appearance of life can be divided into three steps:

1. The first organic molecules—amino acids, sugars, nucleic bases—are formed from simple chemicals such as ammonia and methane.
2. A set of interrelated reactions between these organic molecules appears, and with it the first metabolism. The formation of the first macromolecules, through the stringing together of simple components, makes it possible to catalyze the chemical reactions forming the metabolism; at the same time, these macromolecules—nucleic acids—gradually acquire the power of self-replication.
3. The isolation of these early chemical systems, probably by means of a membrane, leads to the formation of the first self-reproducing cell.

Despite the best efforts of biology and chemistry textbook authors, the vagueness and inherent weaknesses of this account are impossible to hide. First, it does not say what the first macromolecules were, nor precisely which reactions they catalyzed. Nor does it explain why the self-replicative power of these macromolecules should have appeared simultaneously with their metabolic function. Additionally, it fails to give due consideration to the fact that isolation by means of a membrane, though it favors the existence of chemical reactions within a cell, also limits access to external organic molecules. Last but not least, it obscures the fact that cell division is by no means a simple process.

Unsatisfactory though it is, this scenario is nevertheless useful in helping to orient our review of studies of prebiotic chemistry. The first step, which in the wake of Miller's pioneering work no longer seemed the most difficult, is currently a main focus of research. Amino acids and sugars form with relative ease in the sort of conditions that are thought to have existed at the time the Earth was formed. Over the past few years it has been shown that these elementary components are present (indeed they are abundant) in the extraterrestrial environment, in interstellar dust, meteorites, and comets—hence the idea that bombardment by such bodies helped produce fairly high levels of these substances on the Earth's surface. This is not a return to the theory of panspermia: the claim is not that life on Earth originated on another planet, only that the fundamental molecules of life as we know it could have come from the extraterrestrial environment.

Three difficulties in this first stage of the account still remain. First, from all the candidate molecules that form either spontaneously or otherwise in a uncomplicated way, terrestrial life has chosen ones that display certain shapes rather than others—in particular, amino acids having an "L" shape and sugars having a "D" shape. First noticed in the mid-nineteenth century, these shapes stimulated a great deal of research and even more speculation. Using magnetic fields, polarized light, and special surfaces, Pasteur and other scientists tried to re-create an environment that was congenial to their formation, on the assumption that the internal environment of a cell was such that in the course of certain chemical reactions it preferentially induced the incorporation of molecules having a particular form. (We know today that the preference of organisms for particular molecular forms is not so strict: D-shaped amino acids are found in the outer membranes of bacteria and in certain peptides produced by multicellular organisms.) No evolutionary advantage seems to have been conferred by the affinity for L-amino acids and D-sugars; it appears to be merely a sign of the specificity of the chemical reactions that are characteristic of life. Why these shapes should have been chosen in the first place remains puzzling, however. Current theories imagine an initial, weak disequilibrium between the two forms—originating in the conditions in which abiotic reactions took place or in the possible sequestration of one or the other form, for instance by a clay surface—followed by an amplification of this initial disequilibrium by the increasing selectivity of the metabolic reactions.[1]

The second problem is more serious. How could metabolic pathways originate in a hostile environment containing so many similar, though not identical organic molecules? Although the discovery of a variety of organic molecules in the prebiotic environment came as a happy surprise, the set of prebiotic molecules turns out not to have been identical with the set known to be present in organisms, and the ways in which early prebiotic systems were able to select their components are utterly ignored by the standard account.

The third difficulty specifically concerns nucleic acids. Whatever initial conditions are assumed, nucleotides (the building blocks of nucleic acids) cannot be synthesized in a test tube. It is possible, though difficult, to artificially create certain bases (which constitute one of the three components of nucleotides, together with a sugar—ribose or deoxyribose—and a phosphate group), but they do not link up with sugars in the same way as they do in cells. Furthermore, neither ribose nor deoxyribose forms in large quantities in vitro.[2] There remains a gaping hole in our knowledge of prebiotic chemistry. Some researchers argue that these observations lend support to the hypothesis that not only the living world we know, but also the RNA world, were preceded by other worlds: other sugars (such as threose) and nucleotides could have been involved.[3] But it may also be that very particular environmental conditions (levels of temperature and salt concentration) facilitated the creation of the bases and nucleotides found in organisms.

At the moment researchers are engaged in a lively argument over whether life appeared on the surface of the Earth in the primeval "soup" assumed by the standard account, warmed by the sun's rays and enriched in its basic components by incessant meteorite strikes, or in deep-sea springs, whose rich mineral reserves made them a more favorable environment for the development of a primitive metabolism. On the latter view, the organic components that had accumulated in the primeval soup played no role in the appearance of life, and the first stage of the standard account, as I have just described it, simply never happened.[4] The German biochemist Günter Wächtershäuser has argued that prebiotic chemistry began by exploiting the oxidoreductive potential of sulphurous ion derivatives present in these deep-sea springs.[5] More recently the American organic geochemist George Cody has proposed an ecumenical model that combines elements from submarine hot springs and the primordial soup in places located in close proximity to both, along continental margins.[6] There is no consensus yet regarding either the place where life emerged or the nature of the earliest metabolic reactions.

The second step in the received theory involves the formation of macromolecules through the linking together of elementary molecules; the appearance of a metabolism, which is to say at least some of the chemical reactions characteristically associated with life that permit organisms to synthesize their component parts; and the formation of self-replicative macromolecules. The metabolic reactions are necessary for the formation of macromolecular catalysts, which act in turn to accelerate them. What needs to be explained here is how an autopoietic system capable of reproducing its functional components—or at least those components that were hard to find in the external environment—came into existence.

Studies of this second stage of biogenesis have been quite varied, reflecting the many (and often contradictory) opinions of what constitutes the basis of life. Most researchers believe that the first organisms found many of the resources they needed in the soup around them. If so, their metabolism might have been simpler than that of organisms today. Similarly, it is possible that the first catalysts were not macromolecules, but inorganic compounds such as metals, or else simple molecules like amino acids or amino-acid-derived molecules. Some researchers have even suggested that mineral surfaces could have served this function, by concentrating and orienting molecules.

A great deal of work has also been devoted to the problem of macromolecular replication. Today's living world mobilizes a highly complex molecular machinery for this purpose (protein replication involves several nucleic acids and nucleic acid replication involves proteins). Since this machinery did not exist in the prebiotic world, there must have been a simpler way of replicating macromolecules. Because of the linear structure of nucleic acids and the specificity of their complementary base sequences, self-replication is easier to imagine in the case of a nucleic acid macromolecule, such as DNA or RNA, than of a protein macromolecule. Self-replication need not have been direct: in the case of nucleic acids, it is more plausible to suppose that a complementary copy was generated and then copied again to recreate the original molecule. Many research teams have tried to create an RNA molecule that can accomplish self-replication either directly or indirectly. Small advances have so far been made through the creation of RNA molecules that are able to create a bond between oligonucleotide molecules and, more recently, to specifically associate nucleotides with a chain of RNA.[7]

Let us suppose that such "replicases" can in fact be produced. In the endless cycles of replication that would ensue, inexact copies of these replicases would be produced. Mathematical models show that if these inexact copies are more effective, faster-acting, or more faithful than the original forms, they will tend to replace earlier versions.[8] An optimization process (often somewhat hastily called "Darwinian") will then be set in motion.[9]

There is no major obstacle to obtaining RNA molecules having the different catalytic activities required by the primordial cell. A great many in-vitro evolution experiments have yielded RNAs capable of catalyzing most, if not all, of the reactions catalyzed by proteins (including an RNA that forms the C–C bond during a crucial stage of the energetic metabolism, the aldol reaction).[10] The key difficulty lies in linking the self-replicatory properties with the catalytic and structural functions fulfilled by such macromolecules. The idea of a coupling process between different systems of selection to generate a hypercycle, proposed by the German biochemist and physicist Manfred Eigen, is brilliant but highly abstract.[11] We need to be able to precisely explain, for example, how self-replicating macromolecules could also acquire the ability to catalyze the formation of the base elements out of which they are made, and what advantage they would confer on any cell containing them.

In an RNA world, it is theoretically possible that a single kind of macromolecule might be able both to carry out catalytic functions—thus participating in the creation of a prebiotic metabolism—and to produce a copy of itself—thus transferring information to another macromolecule. This still does not solve the problem of exactly how such a sequence of events came about in the terrestrial world, however.

Hovering in the background of this research are the two different conceptions of life that I discussed earlier: life as a self-sustaining chemical system and life as a self-reproducing macromolecular system.[12] In both cases the genesis of life consists in the formation of self-organized systems, but the nature of the components differs: small molecules suffice in the case of a self-sustaining system, whereas informational macromolecules are required for self-reproduction. The first view is inspired by biochemistry, the second by genetics. Each of these approaches seeks to describe the formation of an organism having both characteristics, but each supposes that one of these characteristics, being more fundamental, preceded the other. Both approaches face the same problem of explaining how the second property of life become associated with the first one.

One might imagine that the appearance of life was associated with the creation of a network of proteins in a transsynthetic relation, whereby each macromolecule was involved in one or more steps of the biosynthesis of the other macromolecules. In such a schema—a sort of hybrid of the two views I have just mentioned—macromolecular self-replication is assumed to have been the result of a macromolecular metabolism that appeared before any other form of chemical transformation.[13] Even on this hypothesis, however, one would still have to explain how a metabolism of molecular components capable of forming such macromolecules could subsequently be united with an interacting network of macromolecules.

We come, finally, to the third step in the standard account of the formation of life, the appearance of autonomous living systems. In all known life forms today, biological autonomy depends on the existence of a membrane. At the moment few researchers (with the notable exceptions of the American biochemist David Deamer, the Italian Pier Luigi Luisi, and most recently the American Jack Szostak) are studying the formation of membranes and their fundamental components, lipids. Part of the reason for this is that because the chemistry involved is both unusual and difficult, it has attracted less interest than the chemistry of nucleotides or amino acids; also that lipids (and sugars), associated with earlier work in biochemistry, were far less fashionable than nucleic acids and proteins at a time when molecular biology was enjoying great success.[14] As a result, we know virtually nothing about the prebiotic existence of lipids.

Furthermore, there is no general agreement about the importance of the creation of cell membranes in biogenesis. For some scientists it is the key step, the development that certifies the existence of life. Others take the position that, although some kind of isolation from the external environment would have been necessary for competitive processes to be set in motion, and along with them the evolution of prebiotic forms, mechanisms other than enclosure by a membrane are readily imaginable.

In addition to disagreement over the exact role of membranic isolation in the formation of life, there is also the fundamental problem that any such activity would have had to be selective, allowing small molecules (amino acids, sugars, and nucleotides) to enter freely while blocking the passage of large self-replicating macromolecules. This kind of selectivity is indeed characteristic of the cell membranes we are familiar with, but it is due to the presence within the membrane of proteins that serve as a sort of sieve, permitting small molecules to pass in and out of the cell. Primitive membranes, being formed of lipids that differed from the ones that constitute present-day membranes, were probably more permeable.

I have deliberately refrained from giving a detailed account of recent research on biogenesis, which can be found in a great many books and articles.[15] It includes studies aimed at uncovering the details of metabolic evolution, such as the appearance of organisms capable of capturing the light energy of the sun and subsequently using it to extract oxygen from water; and others that seek to explain how an RNA world led on to a world of DNA and proteins, and how base sequences came to be related to particular amino acids—in other words, how the genetic code appeared. But even a very rapid overview is enough to show that the many different approaches to the problem do not yet converge. Researchers agree neither on the precise order of the essential steps in the genesis of life nor on their very nature. What is more, every line of inquiry runs up against apparently insuperable obstacles.

The difficulty is not so much how each particular piece of the puzzle appeared, but rather how the various pieces fit together. For example, it is relatively simple to imagine how a membrane could form a vesicle; it is harder, however, to see how the formation of the vesicle assisted chemical reactions and favored the self-replication of the macromolecules enclosed within it. Szostak's beautiful demonstration of a direct physical connection between the growth of lipid vesicles and the quantity of nucleic acids they contain suggests a possible link between self-replication of informational macromolecules and the presumptive advantage they confer upon vesicles in which they are present. This link has yet to be established, however.[16] Similarly, it is unclear how a relation came to be established between the first self-replicating nucleotide-based oligomers (if indeed they existed in this form) and the biochemical reactions that were taking place around them.

It is well to keep in mind that these problems, which break down into many smaller problems, none of which seems impossibly difficult, are not so different from the challenge that faced the early molecular biologists once they had identified the origin of life with the invention of the genetic code, namely, to explain how an organism could appear that used the information contained in DNA in order to make proteins, but that at the same time needed proteins to read and replicate this information. For science to advance, researchers must have some degree of confidence that current difficulties, no matter how great they may appear, will one day be resolved—that the solution lies just around the next corner. It sometimes helps to be able to forget such difficulties for a while; indeed, many important scientific results have been achieved by outsiders who had the advantage of not appreciating the magnitude of the problem they set out to solve.

Authors claiming to summarize the current state of knowledge about the origin of life too often tend to minimize these difficulties, however, even to the point of acting as though the problem were already solved and a sequence of events well established.[17] But in order to conclude that life did in fact originate according to some series of individual steps, without violating the geological principle of uniformitarianism (which asserts that the Earth for most of its history has been shaped by slow processes similar to the ones we see at work today), it needs to be shown that these steps could have occurred under similar environmental conditions. This has not yet been done.[18] In order for the merely possible to be converted into a reality, one has only to disregard the less probable assumptions of a given computer model, which typically posit the existence of objects possessing remarkable properties the like of which no object in the real world has, and to neglect to examine the model's many parameters, which can be set as the modeler wishes, frequently, and more or less explicitly, in such a way that the model works.

Moreover, owing to the ambiguity that infects many terms, using a word in two different ways in rapid succession makes it possible to conceal or ignore the actual difficulties involved in passing from one step to another. The best example of this, as we have seen, is the use of the terms "genetic" and "informational." Describing a nucleic acid molecule as intrinsically informational or genetic is a way of brushing aside the problem of how self-replicating molecules were able to acquire additional functions distinct from simple self-renewal. It also often happens that ad hoc hypotheses are invoked. The structure of mineral surfaces, for example, is sometimes said to have an "organizational capacity" that accounts for the emergence of metabolism, the replication of the first RNA molecules, and the formation of vesicles. No matter how interesting the structure of these surfaces may be, and despite some preliminary experiments, the imputation to it of a certain notional capability hardly constitutes a satisfactory explanation of what is, after all, a very complicated phenomenon.[19]

This way of evading difficulties, whether consciously or otherwise, is evidence of the very same unreasoning behavior that scientists rightly criticize in the opponents of science. There are few more perverse forms of irrationality than the one that hides itself behind a mask of rationality, glossing over the contradictions of an argument and refusing to see the holes in it. In the long run, failure to face up to genuine problems undermines the credibility of science and, more generally, the very possibility of rational knowledge itself.

Faced with even the simplest present-day cell, a bacterium, with its thousands of exquisitely catalyzed and regulated metabolic reactions, the enormity of the task of explaining the natural causes of such a wonderful chemical machine can only make scientists acutely aware of the fact that for a long while to come their theories will remain extremely simplified versions of reality. But the scale of the difficulties they face must not stop them from trying. Researchers who tackle the problem of how the first cell came into existence—a cell that obviously was far less complex than its highly evolved modern descendants—can take heart from the impressive advances in our understanding of the origins of life that have been made during the past half-century. However quickly or slowly progress may be made in the future, it is clear that the nature and origin of life are now wholly respectable subjects for scientific investigation.

8

Reading the Palimpsest of Life

In this chapter and the one following I take up the third line of research I mentioned earlier. It flows from a single assumption, namely, that in today's world there exist organisms that are wholly or partly identical to the first forms of life—what are sometimes called living fossils. Microorganisms that are indistinguishable from ancestral forms would be the microbial equivalent of the coelacanth, a fish well known from the fossil record that was thought to have been extinct for many millions of years and then, in the 1930s, was found to be alive in the depths of the Indian Ocean around the Comores (more recently it has also been discovered off the coast of Indonesia). There have been many pseudo-discoveries of living microfossils, among them *Bathibius haeckelii,* a kind of primitive slime identified by the English biologist T. H. Huxley at the end of the nineteenth century, and the more recent description of *Kakabekia umbellata* in the 1960s, but no such findings have survived closer examination.[1]

In the following chapter I shall discuss current research on hyperthermophilic archaea. To begin with, however, we need to consider the hypothesis that every living organism is a living fossil in the sense that one or more pieces of its cellular machinery bear the characteristic traces of the earliest life forms, which subsequent evolution has not entirely eradicated. From this point of view, organisms are *palimpsests* of life. A palimpsest is a parchment that has been written on over and over again: with each new inscription the previous text has been erased, but not completely, so that it is possible to detect beneath one text the traces of another and, in some cases, to reconstruct it. Today's life forms, which use DNA as a memory molecule and proteins as active agents, are supposed by analogy to retain traces of previous living worlds, such as an RNA world, and traces of the prebiotic world.[2] In principle these traces can be recognized and the path followed by life during evolution reconstructed, but one needs to be sure that they are genuine ancestral traces and not later adaptations. The methodological principle that guides such studies is parsimony: the most likely ancestral state is that which would lead to the current diversity of the living world in as few steps as possible. Cladistics, a technique widely used in research on phylogeny, is based on the same principle.

Matters are relatively simple when all contemporary organisms have a particular characteristic and this is quite evidently ancestral. An excellent example is the fact that in every living cell, even today, RNA catalyzes the formation of bonds between amino acids. Proteins are synthesized on a large particle—a ribosome—made of RNA and a variety of proteins. What is more, amino acid bonds are formed at a precise site on this structure. Images recently obtained by X-ray diffraction studies of ribosomal crystals clearly show that this so-called active site contains only RNA, and many experiments performed since have confirmed the role of ribosomal RNA in protein synthesis.[3] A better proof of the hypothesis that an RNA world preceded the present-day living world of DNA and proteins is difficult to imagine: the protein being synthesized in your cells as you read this is being produced by RNA, thus repeating the transition from one world to the other.

Yet this magnificent example should not blind us to the risks involved in such efforts to reconstitute the past. In the early 1980s, the hypothesis of an initial RNA world was introduced following the discovery of intron "self-splicing"—the process by which an RNA molecule copied from a DNA molecule is able to eliminate the noncoding parts of its sequence. It was the first catalytic function to be clearly associated with an RNA molecule. Today most biologists agree that the variety of mechanisms by which the splicing process is currently carried out, and their distribution among modern organisms, suggest that most processes of intron splicing are in fact relatively recent evolutionary inventions. Ironically, then, the existence of an RNA world was asserted on the basis of a phenomenon that we now have strong reason to believe was not part of that world.

We must therefore be careful in reading the palimpsest of life. A number of other observations either confirm the hypothesis of a first living world without proteins or provide insights into the kinds of chemical reactions that primitive organisms were able to carry out. For example, all contemporary life forms make use of various small molecules having a specific structure, often derived from nucleotides, that act as coenzymes; that is, they join with proteins in carrying out metabolic chemical reactions. These compounds make up a large number of vitamins: complex organisms (such as human beings) that have lost the ability to synthesize them are obliged to acquire them through nutrition. It seems probable that these coenzymes are the traces of a primitive, protein-free form of catalysis. The fact that some of them can still directly interact with messenger RNA and control its translation reinforces this hypothesis.[4] Similarly, the major role that metallic ions such as iron, molybdenum, nickel, and manganese continue to play in catalyzed organic reactions, together with proteins that use these ions as coenzymes, furnishes strong evidence that these ions participated in early biochemical reactions as well, and, in the case of iron, lends support to the scenarios for the appearance of life that have been proposed by Günter Wächtershäuser.[5]

Another interesting observation was reported recently by the Spanish biochemist Chomin Cunchillos and his French colleague Guillaume Lecointre, who have studied metabolic pathways using the cladistic principles developed to situate organisms in the evolutionary tree, with the purpose of creating a sound basis for the evolutionary history of metabolism.[6] Their initial results indicate that the degradation of amino acids was one of the first metabolic steps developed by organisms, which agrees with what we know about the abundance of these compounds in the prebiotic environment.

Finally, the analysis of the reactions involved in nucleotide synthesis, showing that the specific forms of these molecules, found in DNA, are derived from ones found in RNA, provides further confirmation—this time at the metabolic level—of the precedence of RNA over DNA. The complexity of the enzymatic reactions involved in the synthesis of the components of DNA (which are beyond the power of even the most efficient catalytic RNA molecules to achieve) and the fact that the machinery involved in copying DNA into RNA is composed only of proteins (whereas protein synthesis mainly involves RNA) also strongly suggests that DNA was the last of the three kinds of informational macromolecules of life to appear, coming after both RNA and proteins.[7]

For the moment we can read no more of the palimpsest of life than this. Many characteristics of today's organisms tell us nothing about what led to their appearance. Consider, for example, the genetic code and the twenty amino acids that have been selected by evolution to form proteins. How did the genetic code appear? Was it preceded by a simpler code, using only two nucleotides, that permitted the synthesis of proteins containing only a small range of amino acids? Despite a substantial amount of research, and numerous observations indicating that an evolutionary optimization process has minimized the effects of coding mistakes, not one of the many hypotheses that have been put forward has been proven.[8]

Nor should the originality of recent speculation be overestimated. The nineteenth-century German naturalist Ernst Haeckel's now discredited principle of recapitulation, according to which each organism during the course of its embryological development exhibits all the stages of the evolutionary process that led up to it ("ontogeny recapitulates phylogeny"), was based on the same idea: each organism carries within it traces of the history of life. What is new, however, is that this idea is now being applied to the fundamental molecular mechanisms of living organisms, and extended to the stages that preceded life itself.

Although attempts to read the palimpsest of life have, for the time being at least, yielded relatively modest results, they nevertheless show that metabolism, or chemical transformation, is essential to life. In any case it is plain that the biological parchment that has come down to us bears many more traces of the elements of a primeval soup, hydrothermal springs, and catalytic clays than of informational molecules or autonomous replicators. There is one approach, however, that promises to give some insight into the characteristics of extinct organisms (or microorganisms) and the environment in which they were living. Each DNA sequence is in some sense a palimpsest as well: by comparing a given sequence with sequences of organisms belonging to other species, molecular phylogenies can be drawn up from which the gene sequence of the common ancestor of the different species being studied can then more or less accurately be reconstructed. From this sequence it then becomes possible to synthesize the corresponding ancestral protein—a protein that no longer exists, but whose biochemical characteristics help to illuminate the conditions in which the ancestral organism developed.

The ambition is the same as in *Jurassic Park*, but because the reconstitution concerns only individual proteins the problem is more tractable. By applying this strategy to a universal protein involved in protein synthesis, the American biochemist Steven Benner and his colleagues have proposed that the common ancestor to all eubacteria was a moderate thermophile living within a temperature range of 55–65°C.[9] This is a result of considerable interest for the ongoing debate over the place of hyperthermophilic archaea in the early stages of life, to which I now turn.

9

Life Under Extreme Conditions

It has long been known that organisms can resist extreme conditions. One has only to think of hibernation, or of the formation of spores, discovered by microbiologists toward the end of the nineteenth century. Recent studies have succeeded in describing some of the mechanisms involved. They have also pushed back the limits of life still further by demonstrating, for example, the abundance of organic forms found in ice and the resistance shown by some organisms to very high pressure.[1]

The study of hyperthermophilic extremophiles—organisms that preferentially develop at very high temperatures—has now become a very active field of modern biology. The first reason for this is that many of these organisms have only recently been discovered. When Carl Woese suggested that the living world is formed of not two branches, but three (eukaryotic organisms, eubacteria, and archaea), many of the first archaea to be characterized were hyperthermophiles. The very term first used by Woese, "archaebacteria," indicates that these organisms were initially thought to be primitive bacteria, closer to aboriginal organisms than modern eubacteria, which were considered more highly evolved. Shortly afterward it was discovered that hitherto unknown hyperthermophilic archaea flourish in the depths of the ocean, many thousands of feet down, where the Earth's tectonic plates separate and release magma, creating massive hot springs that act as a breeding ground for these organisms. A new world of life had come into view, awakened from eternal night by the searchlights of submarine explorers.

Owing to their special metabolism and their ability to develop at very high temperatures, contemporary hyperthermophilic organisms are of great interest—not least because enzymes useful for biotechnological purposes can be extracted from them. The polymerase chain reaction, a widely used technique for amplifying DNA that is at the heart of a great many modern diagnostic advances and much contemporary research in genetic engineering, exploits the ability of these enzymes to resist cycles of intense heating and cooling. There is almost unanimous agreement today that the study of such organisms can also tell us a great deal about whether or not extraterrestrial life forms exist. Any planets and moons on which we may hope to find organisms are likely to have a harsher climate than that of Earth. The discovery and study of extremophile organisms has made it possible to extend the temperature and pressure conditions under which until recently life was thought to be capable of developing, although, it must be said, rather less than is generally suggested.

Yet another reason for the current interest in these organisms, one that is less clearly stated but nonetheless widely shared, is that extremophiles—in particular, hyperthermophiles—may resemble the first forms of life that appeared on Earth. If one assumes that life did not in fact develop in a warm primeval soup, but rather in the depths of the oceans, near hot thermal springs, this seems a reasonable enough hypothesis. Indeed, the idea has found acceptance among a far broader circle of scientists than adherents of the thermal spring theory, probably because the difficulties facing organisms in a hostile terrestrial environment, exposed to the ravages of a disordered climate and the sun's ultraviolet rays, unfiltered by the Earth's primitive atmosphere, seem evident and in accord with our intuitive conception of the earliest phase of the organic world. Nonetheless, the theory stands in need of critical examination, for the opposite view—that all contemporary life forms, including extremophiles, are the product of long evolution and adaptation—is equally probable, if not more so.[2]

After all, one hardly wishes to find oneself in the awkward position of discovering that a characteristic considered fundamental to life, and one of its ancestral traits, is in fact a recently evolved characteristic. Belief in the stability of genetic material, for example, was one of the hallmarks of the genetic view of life that developed from the end of the nineteenth century onward; indeed, genes were partly defined by their supposed permanence.[3] Over the past three decades, however, many studies have shown that this durability is due in part to the chemical composition of the DNA molecule, whose superior resistance to changes in the environment no doubt encouraged the passage from an RNA world to a DNA world, but mainly to the existence of a great many particularly complex repair mechanisms that continually correct errors which appear during duplication or are caused by environmentally induced alterations. The perfection of these repair mechanisms, over hundreds of millions of years of evolution, has conferred upon DNA a far greater degree of stability than the fact that it is chemically inert would lead one to expect.

The same is true of extremophile organisms. Their special character is the product of complex molecular mechanisms that progressively appeared during the course of evolution. One of the clearest examples of this is the ability of the DNA molecule found in these organisms to withstand very high temperatures. The DNA double helix normally reacts to a rise in temperature by unraveling into two separate strands. To limit the effects of this denaturing process, hyperthermophilic bacteria roll up their DNA, thus reducing the possibility that the DNA molecule will disentangle. This operation is carried out by a specialized enzyme, reverse gyrase, an evolutionary invention that exists only in hyperthermophiles, enabling them to adapt to extreme environments.[4] We need therefore to abandon the idea that such environments are the normal milieu for life: severe heat or cold, high salinity, drought, high acidity, and high alkalinity are all unfavorable conditions, to which organisms have been able to adapt only by evolving solutions that often resemble ones devised by human beings to cope with similar conditions. Fish and bacteria that live in very cold water, for example, synthesize a kind of glycerol that, like the antifreeze used in cars to protect engine coolant from the effects of cold weather, lowers the freezing point of water and retards the formation of ice crystals.

Probably the most interesting research from the point of view of trying to discover the nature of life is concerned with examples of what might be called suspended life: spores that are formed from microorganisms in the face of a suddenly hostile environment, organisms that can remain in a desiccated state for years or even decades, or microorganisms living miles beneath the surface of the Earth that have recently been discovered to have a very low rate of metabolic activity. Life can also be suspended by artificial procedures such as immersion in liquid nitrogen, a technique developed in the 1920s and now widely used to conserve cells, sperm, eggs, and embryos. In all these cases, where life has been stopped but not irretrievably lost, two characteristics of living organisms that are normally closely linked with each other—metabolic activity (that is, the continual renewal of the molecules that make up the organism) and structural organization—have been dissociated, the former being interrupted and the latter maintained. Because the various means by which organisms are able to endure extreme conditions, whether natural or artificial, have the effect of preserving their cellular structure, life is able to do without the chemical activity that is also one of its distinguishing features, at least temporarily.

Does this imply that among the several characteristics closely associated with life, structural organization is primary and chemical (or metabolic) activity secondary? Once again we need to be extremely careful in drawing conclusions. It is true that the hereditary organization of contemporary life forms enables them to survive a temporary interruption of chemical activity. But that does not necessarily mean that the first organisms were also able to perform this trick—far from it. It is more probable that resistance to such metabolic interruptions is the result of the evolutionary creation of mechanisms that make it possible to stabilize cellular structure and to repair the damage that typically is caused by a return to normal metabolic activity.

In the case of a suspension of life through immersion in liquid nitrogen, it might be objected that this is a wholly artificial situation to which organisms could not have adapted in the course of evolution, and that for this reason it demonstrates the primacy of structural organization. It needs to be kept in mind, however, that although this method of arresting chemical activity is indeed unprecedented, the possibility is not therefore excluded that the means by which the organism compensates for the absence of such activity under unnatural circumstances may in fact be the result of an adaptation to other quite natural conditions in which the metabolism slows down. To take an opposite example: it sometimes happens that life is able to sustain itself by metabolic activity despite the fact that its structure has been profoundly modified and its power to reproduce utterly abolished. One thinks of red blood cells, which transport oxygen in the blood. These are cells with an active metabolism that have lost their nucleus and, as a result, no longer contain the informational image of themselves that all other cells possess. The life span of red blood cells is limited, but during this period they are fully alive.

The study of life under extreme conditions therefore has little to teach us about either the origin or the nature of life itself. It simply shows that organisms have learned to resist and to adapt to conditions that would otherwise be fatal to the continuation of life. Research on extremophiles indicates how far life has been able to expand the limits of its endurance, starting from what must have been very low chances of survival. But it decides nothing with regard to the probability that life appeared and evolved under extreme conditions. Another, bolder line of research aims to manipulate the genetic code of an organism or to enlarge the number of amino acids available to it (or to introduce some ambiguity in the code), either by means of targeted modifications or otherwise by exerting a strong selection pressure upon a broad population of organisms.[5] Apart from its potential biotechnological interest, this approach is notable for its assumption that by pushing back the frontiers of life still further we will come to have a better understanding of its fundamental nature, just as a test pilot learns more about an airplane by taking it up to and beyond its limits. The same is true on a personal level: by seeking to extend our physical and mental capacities we discover more about ourselves.

Even so, some of the wilder claims made by researchers hoping to be able to create entirely new life forms, ones having a different molecular makeup than present-day organisms, need to be discounted. Adding one or two amino acids to the list of those that are already used on Earth would not be revolutionary. Many organisms, as we are now gradually discovering, have done precisely this during the course of their evolution.[6] We need also to be mindful of the fact that the functional organization of any new life forms that may be created in this way would be based on that of existing organisms, and to that extent they would not be entirely new. This makes it all the more unlikely that such studies will be of much help to us in our search for an answer to the question “What is life?”

10

The Search for a Minimal Genome

In the opening chapters I discussed the ways in which the conception of life has changed over the past century, and emphasized the importance initially attached to viruses before it was recognized that they are in fact parasites, which fatally undermined the hopes that had been placed in them as a model system. The idea that the nature of vital properties can be understood by studying the simplest organisms is a legacy of Lamarck, and one that, unlike his theories of variation and evolution, is still widely accepted today. The advantage of such organisms, Lamarck argued, is that they present "all the conditions necessary to the existence of life and nothing else beyond."[1]

This view led to the search for an elementary, formless organism, whose existence was mistakenly supposed to have been demonstrated in the late nineteenth century. Despite many such failures, confidence that the nature and origin of life will one day be understood by studying the simplest organisms remains unshaken. Indeed, it has been reinforced by the fact that it was the study of a humble bacterium, *Escherichia coli,* that led to the spectacular growth of molecular biology in the second half of the twentieth century, revealing the fundamental principles of the macromolecular organization of the entire living world.[2] The recent discovery, in 2002, of a hyperthermophilic archaebacterium (or "nanobacterium," so called because of its exceedingly small size, very different from all known archaea) attracted great attention and rekindled hopes that an organism ideally suited for studying the mysteries of life had at last been discovered.[3] Argument continues to rage over the actual existence of the microscopic life forms that the Finnish physician and microbiologist Olavi Kajander claimed to have discovered a few years earlier.[4]

For a geneticist, a simple organism is one with a limited number of genes. The study of such organisms has taken a new turn in the past decade with the genomic sequencing of an intracellular bacterial parasite, *Mycoplasma genitalium* (associated in humans with genital infections, as its name suggests).[5] On discovering that the genome of this bacterium contains only 500 genes, one-seventh of the number found in *E. coli,* researchers immediately seized the opportunity it presented for determining the minimal genome, which is to say the smallest genome that is compatible with life. The American scientific entrepreneur Craig Venter, well known for his pioneering work on the human genome project and the leader of the group that sequenced *Mycoplasma genitalium,* has set the ambitious goal of synthesizing from scratch a completely artificial organism that carries out only those functions which are necessary and sufficient for life. This, as Venter acknowledges, is one way of trying to answer the question "What is life?"[6]

Two approaches have been pursued in parallel. The first aims to further simplify the genome of *Mycoplasma genitalium* by removing seemingly useless, or nonessential, genes through mutagenesis.[7] A step-by-step procedure succeeded in showing that up to 150 genes could be removed, one at a time, with no apparent ill effects. Note, however, that this does not prove that all 150 genes could be removed at the same time.

A second approach proceeds from the idea that the unique properties of an organism must represent a kind of nonessential addition. Accordingly, if the genes that are common to all organisms could be identified, one would have discovered which ones are indispensable for life. This amounts, then, to looking for the lowest common genetic denominator. A small genome like that of *Mycoplasma genitalium* is a good place to start, because nature has already done much of the work. Preliminary research suggests that the smallest possible number of genes may be 250 or so.[8] It has been supplemented by the sequencing of other small genomes, among them the genome of a bacterial insect symbiont that is phylogenetically close to *E. coli*, which makes it easier to analyze, and the evolutionary vestige of the nucleus of a eukaryotic cell, known as a nucleomorph, with only 331 genes.[9]

What can we learn from these results? The genes present in organisms with small genomes code for enzymes involved in both the replication of DNA and the associated processes of transcription and translation, as well as for proteins involved in intracellular transport and so on—functions that we already know to be fundamental in all living cells.[10] The question arises how these organisms came to have minimal amounts of genetic information in the first place. Most of them are parasites or symbionts that take advantage of host organisms. The genes these organisms lack are mainly ones responsible for metabolism and the synthesis of amino acids, nucleic bases, vitamins, and membrane components—functions that the parasite has managed during the course of development to borrow from its host. In this respect small genomes are an example of what André Lwoff once called "regressive evolution."[11]

The study of such small-scale life forms today, like that of viruses more than fifty years ago, leads to a dead end: they are simple only because other organisms do their work for them. The discoverers of the hyperthermophilic nanobacterium—a parasite of another archaebacterium—implicitly admit as much in maintaining that this organism occupies an "intermediate" place between viruses and the smallest known organisms.[12] More generally, however, the question arises whether it may perhaps be a mistake to try to answer the question "What is life?" by studying any given organism.[13] After all, life is a planetary phenomenon, a worldwide web of mutually dependent and interacting organisms. In today's living world an individual organism survives only because it makes use of the work done by other organisms; in return, it makes its own contribution to the survival of its fellow organisms, which jointly form a closed network. In the last chapter of this section I shall examine the view of life as an ecological system in greater detail.

For the moment we need to recognize that the search for a minimal genome rests on an implicit assumption having no justification whatsoever, namely, that there exists only one form of life. To the contrary, it is very likely that the problem of identifying a minimum set of genes consistent with organic function will admit not of a single solution but of many.[14] What is more, there is no reason to believe that any of these solutions will be the one that nature actually adopted in the case of the first cell containing genetic material. Like the other characteristics of organisms, the minimal genome is a product of evolution, made possible by the development of complex processes of exchange and of parasitism. The minimal genome is a highly evolved form of life; it can tell us nothing about the origins of life.

11

Astrobiological Investigations

Speculation about extraterrestrial life has a very long history, going back at least as far as the ancient Greeks. The idea of the existence—and often the superiority—of celestial beings is a venerable feature of religious thought.[1] Like the Persians in Montesquieu's *Persian Letters* (1721), aliens have often served as a vehicle for the hopes—and the fears—of mankind.[2] It is somewhat surprising, then, that one finds scarcely more interest in the question "What is life?" in the writings of astrobiologists today than in the writings of biologists generally. New experiments are proposed, but rarely supported by any argument that the particular characteristic or property being studied is in fact constitutive of life in the broadest sense—that is, that it is a feature of all forms of life, independently of the local conditions in which they have appeared.[3]

Most astrobiologists seem to be unable to overcome the difficulty raised more than forty years ago by George Simpson: "Either the extraterrestrials will be similar to terrestrial organisms, and their study will be without value, or they will be very different, and will not even be recognized as living."[4] The disinclination to inquire into the nature of life seems all the more peculiar in this case since the ambition of astrobiological research is considerable. If one takes the view that a field of study having only one object cannot be regarded as a genuine science, then biology in its present form is disqualified on the ground that all terrestrial organisms appear to have a single common origin. By discovering new forms of life, astrobiologists seek to transform biology into a true science.

For the most part this lack of curiosity regarding the nature of life is explained quite simply: the characteristics being sought are those of terrestrial life, which, because they are supposed to be known to everyone, are in no need of being stated explicitly. Some (though not many) researchers do go further, arguing that because all known organisms are subject to the same chemical constraints—they are formed mainly of water, with a carbon-based chemistry involving amino acids (the simplest compounds that are readily formed from carbon) and macromolecules (the only molecules capable of containing information)—any forms of life that we may find elsewhere are bound to closely resemble terrestrial life forms in their chemical composition and metabolism.[5] But nature might just as easily have found other solutions to the problem of life than the one it devised on Earth. Sugars, for example, have all the characteristics required to form informational macromolecules, despite the fact that they have not done so in our corner of the solar system. This failure of imagination makes astrobiology not so much a new science as the mere adding together of the two disciplines, astronomy and biology, that gave rise to it.

Two British researchers, the biologist Jack Cohen and the mathematician Ian Stewart, have therefore urged the creation of a genuinely new science—xenobiology, or a biology of the strange, as they call it—that would dispose of the charge of "terracentricity" by exploring the nature and limits of life in all its possible forms, not merely ones based on DNA.[6] The intellectual laziness that Cohen and Stewart object to is aggravated by the fact that the sort of answers that are typically given to the question "What is life?" are not necessarily useful for an extraterrestrial research program. If researchers were to adopt an ultra-Darwinian definition ("an organism is an autonomous structure capable of self-reproduction with variation"), for example, it is difficult to see how a space probe, even if it landed on the surface of another planet, could provide experimental evidence for or against the existence of life there. Astrobiological research proposals in any case rarely distinguish between assumptions based on careful reflection about the nature of life and ones that amount merely to a tacit and resigned acceptance of the fact that present-day technology can provide only a limited number of answers.[7]

This kind of difficulty is not restricted to research on extraterrestrial life, no matter that it is probably greater, and more obvious, in this field than in others. There is an old joke about the drunk who loses his car keys and looks for them under a street lamp—not because he thinks he lost them there, but because the light is so much better. Scientific research always takes place at night, with only a few flashlights to illuminate reality. Good researchers adapt the means available to them to the end in view and select the right tool for the right job. In this way they avoid two related dangers: stubbornly pursuing a goal that cannot be attained by means of current experimental techniques, and relying on procedures that are tried and tested, but that cannot provide the desired answers.

The first experiments in exobiology, as the search for extraterrestrial life was originally called, were carried out thirty years ago (more than twenty years before NASA launched its astrobiology program) by two Viking probes sent to Mars. Their mission was to look for organic components on the surface of the planet and to test for the presence of chemical reactions characteristic of the metabolism of extant terrestrial organisms. For the latter purpose, Martian soil samples were analyzed in order to determine their ability to fix carbon dioxide (as most photosynthetic organisms, including plants and some bacteria, are able to do); to modify the composition of the surrounding atmosphere, for example by releasing oxygen (again, like most photosynthetic organisms); and to assimilate and transform a mixture of organic molecules. These experiments assumed that extraterrestrial organisms would share two characteristics in particular with living organisms on Earth: organic composition, and an active metabolism that continually synthesizes complex compounds from, and degrades them into, small molecules such as carbon dioxide or oxygen.

The findings of the two Viking probes were both complicated and ambiguous. On the one hand, no organic compounds were detected either on the Martian surface or immediately beneath it.[8] On the other hand, the three experiments designed to uncover evidence of metabolic activity all yielded positive results, although the Martian observations did not correspond to the ones obtained when the same experiments were conducted on samples of terrestrial soil containing living organisms, before the probes left Earth. More puzzling still, preheating (or "sterilizing") the Martian soil caused the observed reactions to disappear, as though organisms had been eliminated by means of this procedure. Provisionally it was concluded that, in the absence of any organic matter, the observed reactions might only be "pseudo-metabolic" reactions, which is to say the product of other, unknown chemical reactions.[9] A more precise explanation, involving the creation of superoxide ions through the action of ultraviolet light on the surface of Mars, was finally confirmed after almost twenty-five years of experimental activity. It accounts for some (but not all) of the phenomena that had been observed, as well as the vanishingly low frequency on the Martian surface of organic compounds, which, despite the continuous bombardment of the planet by meteorites, are destroyed by these superoxide ions.[10]

The tale of the Viking probes testifies to both the elegance and the quality of the experimental protocols employed, the difficulties encountered in interpreting the results, and the way in which embarrassing data that could not immediately be explained were temporarily swept under the carpet—a kind of voluntary amnesia, in effect. There are a great many successor projects currently under way concerning various celestial bodies, either planets or moons, not all of which belong to our solar system. Some of these projects are based on direct observation from the planetary or lunar surface; others are being carried out by means of long-distance investigation. A useful way of considering these projects for our purposes will be to classify them into two groups, according to what they seek to find.

The first class of projects is not new. It aims not at detecting life itself but at discovering the existence of conditions that are thought to be favorable to life. Investigations into the structure of planets, their surface temperature, the nature and intensity of the rays that bombard their surfaces, and the gaseous composition of their atmosphere can all be regarded as falling under this head. Much of the research is focused on water: either its actual presence today, in liquid or solid form, or traces of past flows of liquid—a sign that, at one time or another, there was water on the body being studied (although the possibility cannot always be excluded that such traces were left by some substance other than water). In our solar system, the main targets of this research are two of Jupiter's moons, Ganymede and, especially, Europa; a moon of Saturn, Enceladus; and the planet Mars. The recent missions to Mars seem to have clearly demonstrated the presence of water on its surface in the distant past, and even relatively recently, though not in any stable way. A consensus is still far from being reached.[11]

Water is generally considered essential to life, and not only to the forms that have developed on Earth. In both marine and land-based life forms an aqueous medium permits the rapid diffusion of molecules, and therefore the chemical reactions that constitute a metabolism. Indeed, it is rather difficult to imagine metabolic activity occurring in a solid. The likelihood that there exists another liquid with the same ability as water to dissolve and concentrate molecules seems very small.[12] And if it is now accepted that solid water—ice—is not completely hostile to life, this is because within sheets of ice there are very often pockets of water. The study of the terrestrial environment has further shown that organisms can accommodate themselves reasonably well to the sort of seasonal freezing that occurs in the Arctic and Antarctic oceans, and to the dry, intensely cold conditions found in the desert valleys of Antarctica.[13]

But the mere presence of water alone is not enough. An aqueous medium has to be able to permit the survival and development of a community of living organisms; that is, it must be able to provide both the organic compounds (whichever ones these may be) and the energy required to ensure the survival, not of a single cell, but of an entire ecosystem. Life is able to emerge only in a stable environment that brings together a sufficient number of interacting organisms.[14]

As we have seen, the study of extremophiles has so far scarcely budged the physical boundaries within which life is thought to be possible. On the other hand, it has succeeded in showing that perfectly viable ecosystems can exist in conditions that were previously considered uninhabitable. Astrobiologists today are therefore less restrictive in defining the kinds of extraterrestrial environment that might harbor life than they were only a few years ago, with the result that the presumptively habitable zone of our galaxy has been enlarged.[15] But in order to be able to nurture life (or to have nurtured it in the past), a planet must have been habitable for a certain period of time, during which its climate must have remained more or less constant. This is true of our own planet, where until recently the climate has changed relatively little. The changes now taking place are due in part to the very existence of life itself and the effects of living organisms upon the Earth's atmosphere.

The simplest objective of this first class of astrobiological projects is to improve our understanding of prebiotic chemistry, which is to say the formation of the building blocks of life in abiotic conditions. This was the purpose of the Cassini-Huygens mission to Titan, a moon of Saturn. Although temperatures on the surface of Titan made the appearance of any life form impossible, they did not prevent the formation of simple organic molecules, the nature of which will be of great interest to astrobiologists.[16] Missions aimed at characterizing the behavior of such molecules in different planetary environments are justified less by any expectation of a major theoretical breakthrough (prebiotic chemistry is rather well understood, at least in comparison with other aspects of the prebiotic process) than by the very possibility of carrying out such missions, which invariably supply fresh empirical data.

A second class of astrobiological research projects is more directly concerned with detecting actual signs of life. We have already seen that an isotopic imbalance may be such a sign. Other signs are more readily identified, such as the presence of oxygen in the atmosphere (the result of photosynthesis by organisms).[17] This, too, is a kind of chemical imbalance. The current concentrations of oxygen in the Earth's atmosphere, for example, can be explained only by the presence of organic chemical reactions that cause it to be continually generated by means of solar energy. (On other planets researchers will be more inclined to look for evidence of ozone, a substance derived from oxygen that is still easier to detect.) Other chemical imbalances present even clearer indications of life. The highly oxidizing character of the Earth's atmosphere, due to the abundance of oxygen in it, would lead one to expect the concentration of a very simple molecule such as methane to be much lower than it is. Atmospheric methane is indeed oxidized but, because it is continually produced by living organisms (in particular, by a group of archaea called methanogens), this loss is offset.

The occurrence of chemical disequilibria in the atmosphere of a planet may therefore be a sign of the presence of organisms. Before adopting such a hypothesis, however, alternative explanations—such as the release of gases from the interior of the planet by volcanoes—have to be carefully considered. The color of a planet may also be a sign of life. If it does not agree with what is known of the composition of the surface, the possibility arises that the color may be due to pigments produced by organisms in order to trap solar energy.

Such pigments are a still more persuasive argument for the existence of life if their abundance (and therefore the color of the planet) varies according to the position of the planet in its orbit, cyclically altering the amount of energy it receives. Simple inspection of planetary landscapes may also help reveal the presence of life. In the case of our own planet, for example, erosion and the transport and deposition of sediments are strongly affected by the presence of organisms.[18]

Just such signs of life were detected by the Galileo space probe in December 1990—on Earth! Following up on early observations made by the Voyager probes, Galileo managed to provide more precise knowledge of Jupiter and its moons, particularly Europa and Ganymede, where the discovery of water and ice raised the possibility that these bodies may be congenial to the development of life. To reduce the amount of time required to reach Jupiter, the Galileo mission planners had accelerated the probe's speed by harnessing the gravitational energy of Earth and Venus, causing it to pass within six hundred miles of our planet a few months after its launch. The probe's onboard observational instruments succeeded in detecting clear signs of terrestrial life, most notably the abundance of oxygen and methane in addition to the characteristic color of its surface, but also radio emissions, some of which were modulated.[19]

The Galileo study should be seen as a sort of control experiment. On the one hand, it shows (or confirms) which signs of life are the easiest to detect. But it also highlights the difficulties facing such missions: given the limited resolution of the probe's observational instruments, no evidence of what is called "technological geometrization"—buildings—could be found. A more systematic search for signs of life through its environmental effects will be initiated in the near future for planets of other solar systems, using instruments based on Earth or in its immediate vicinity. The first step will be to monitor the surface conditions of these planets for evidence that they are hospitable to organisms. The next step will be to look for signs of the presence of life, such as a particular color or the existence of atmospheric chemical imbalances.

More direct methods, similar to the ones employed by the Viking landers, are available to investigate the planets in our own solar system. First, however, it is necessary to locate one or more sites where life might be present. On a planet like Mars, for example, it is possible that although life is not (or is no longer) present on the surface, it may yet be found in subterranean layers. Next, one must choose the right experiments. Looking for signs of metabolic activity is essential; but in evaluating the observational data it will be difficult to discount the possibility that such transformations may be the result of "purely chemical" reactions. The presence of organic compounds (amino acids and sugars, abundant in interstellar dust and on the surface of meteorites) would not be an infallible sign of life unless the predominance of one stereoisomer, or a certain isotopic composition, or evidence of their organization into macromolecular structures were to be found as well.[20]

It is also possible, of course, that life forms can be directly observed, or else fossil traces of them or their mineral remains.[21] The problems involved in analyzing soil samples, and the false hopes that were raised a few years ago following the apparent discovery of fossil traces on a Martian meteorite, only serve to emphasize the caution that must be exercised in interpreting such results. Nevertheless, if life exists, it will very likely have invaded the whole surface of the planet under study, and there will be little trouble in demonstrating its existence, even if the local chemistry differs somewhat from that of life on Earth (assuming, of course, that one of the key features of life is its invasive character, associated with the ability of organisms to reproduce and adapt). It will be far more difficult to adduce proof of extinct life forms, however, and more difficult still to prove that a given planet has never supported life in the past.

There is one last way of looking for life, in the form of signals emanating from organisms that are supposed to have developed a technology equal or superior to that of human beings. The SETI (search for extraterrestrial intelligence) research program uses radiotelescopes to listen for the presence in the universe of "nonnatural" signals, which is to say signals that cannot be explained by our current knowledge of natural phenomena.[22] Despite a great deal of intensive listening (aided by more than four million volunteers working on their desktop computers), no such emissions have been detected yet. This silence may indicate that our means of listening are inappropriate, perhaps because they mistakenly assume that alien civilizations would use the same means of communication as we do (prompting the Oak Ridge Observatory in Harvard, Massachusetts, to begin scanning the sky recently with a powerful new telescope for flashes of light—briefer than a nanosecond—from alien civilizations).[23] Or it may simply mean that there are no intelligent extraterrestrial civilizations, and that the human phenomenon found on Earth is exceedingly rare, perhaps even unique in the universe.

12

Life as a Living System

At first sight, this chapter may appear to be a miscellany of unrelated topics: the structure of eukaryotic cells, death of several kinds, and what genomic sequencing programs can tell us about the mechanisms involved in the evolution of organisms. My purpose, however, is to show—what in this case may be a matter more of reminding ourselves of what we already know—that life cannot be reduced to a mere collection of autonomous organisms.

Every living being is a part of a community that gives organisms their genes and decides whether they live or die, even whether they are living or non-living. At all levels of observation, life is system: from the cell, the building block of multicellular organisms, all the way up to the various natural ecosystems found on Earth. The interrelation of these systems (which François Jacob called "integrons"), whereby each one is a component of another higher-level system, is one of the characteristics of life.[1]

I have already mentioned, more than once, the differences between a bacterial cell and a eukaryotic cell, whose genetic information is enclosed within a nuclear membrane. In addition to a nucleus, most eukaryotic cells contain other intracellular structures known as mitochondria, which use the oxidation of various molecules—mainly sugars, lipids, and amino acids—to produce adenosine triphosphate (ATP) as an energy source. Plant cells have corresponding organelles called chloroplasts, which produce ATP from light energy. Almost a century ago, biologists noted a general similarity between these organelles and bacterial cells, and suggested that the eukaryotic cell was the result of a process of symbiosis, in this case the fusion of previously distinct organisms.

Only since the 1960s, however, under the influence of the American cell biologist Lynn Margulis, has this hypothesis been generally accepted. Two things were decisive in swaying scientific opinion: the discovery that mitochondria and chloroplasts contain their own specific DNA molecules, and the demonstration of many specific molecular similarities between these organelles and bacteria.[2] Although most of the genes that code for the components of mitochondria or chloroplasts are contained in the nuclear chromosome, the traces of an ancient process of endosymbiosis can still be observed. Even so, no modern biologist would suggest that mitochondria or chloroplasts are living entities, though they may once have been alive during the initial phases of this process, more than two billion years ago.

It is possible that the nucleus and the cytoplasm of the eukaryotic cell are themselves the result of one or more symbiotic processes. Unlike the generally accepted view of the endosymbiotic origin of mitochondria and chloroplasts, however, this remains a matter of conjecture. The most recent genomic data suggest an opposite scenario, namely that the nucleus and cytoplasm of the eukaryotic cell derive directly, along with bacteria and archaea, from their last common ancestor, known as "LUCA" or "progenote," which lies at the root of all present living forms. This ancestor is therefore assumed to have been much more complex than its descendants.[3]

In some kinds of symbiosis each partner remains an autonomous organism, which makes it difficult, if not impossible, to define the minimal functions of life. A completed process of endosymbiosis, on the other hand, abolishes the vital character of the organisms involved, to the benefit of the new unit—in this case the eukaryotic cell. This was an important step in the evolution of life; indeed it appears to have been a necessary condition for the formation of complex multicellular organisms, which had nothing to do with the Darwinian model of the evolution by descent with variation. Symbiosis should not be seen as a "hidden dimension" of evolution, however. It frequently leads to a specialization of function, with an associated decrease in the metabolic capacities of the symbionts. Regressive evolution of this sort makes organisms much more fragile. In many cases, in fact, symbiosis represents an evolutionary dead end.[4]

Another way in which evolution by descent with variation can be short-circuited is through the lateral transfer of genes (also known as "horizontal transfer"), a phenomenon that is particularly widespread in the bacterial world, but found also in more complex organisms.[5] Despite a few scattered observations in the scientific literature of the early and mid-twentieth century (Haldane, for example, included such exchanges in his scenarios for the origin of life), the scale of the phenomenon was first fully appreciated only as a result of systematic genome sequencing.[6] In bacteria, lateral gene transfer (and recombination) probably played a major role in adapting organic development to varying conditions.[7] It was crucial in the diffusion of new pathogenic forms, where it rarely involves an isolated gene, but rather a set of genes that immediately gives the recipient organism a selective advantage. Although we now understand some of the mechanisms by which lateral transfer occurs (conjugation and transformation in bacteria, transport by viruses throughout the living world), the exact weight of each mechanism in this process and the possible existence of other, yet unknown mechanisms remain unanswered questions.

Given so much uncertainty, a sense of proportion is called for: we should neither view lateral gene transfer as the sole explanatory mechanism of evolution, which renders all previous observations obsolete, nor turn it into a mere curiosity or even an argument in favor of genetically modified organisms ("If nature does it, so can we"). In most cases it is possible to describe the various forms of a given gene in different species and then to establish a "molecular phylogeny" consisting of evolutionary trees based on descent with variation. In the larger scheme of things, lateral gene transfer is a relatively minor phenomenon.

Biologists nonetheless agree that this phenomenon must have played an important role in the early living world. Eubacteria, primitive archaea, and the precursors of eukaryotic cells probably exchanged genes with one another quite frequently. The stabilization of the genetic code in its present form may have been a consequence of this extensive process of transposition.[8] Together with symbiotic processes, lateral gene transfer had the effect of converting the tree of life into a ring.[9] It is possible, too, that certain evolutionary inventions were rapidly diffused by means of lateral gene transfer: the replacement of RNA by DNA as the physical material of heredity, for example, may have taken place first in viruses before spreading throughout the living world.[10] According to some recent scenarios, the original living world was formed of more or less (probably less) autonomous organisms, among which more than 90 percent of existing genetic material, carried by viruses, circulated freely. On this view, in which a single set of genes, far exceeding the storage capacity of an individual organism, is assumed to have provided the ancestors of modern organisms with their genetic identity, the origin and the earliest phases of the evolution of life did not coincide with the creation of autonomous and self-reproducing life forms.[11]

The potential importance of viruses in the early stages of life (and even in later horizontal gene transfers), emphasized by recent evidence for the richness of the viral world and the discovery of "gigantic" viruses such as the Mimiviruses, should nonetheless not lead us to suppose that viruses constitute a fourth, intermediate kingdom of the living world.[12] However rich may be the genetic information that these viruses bring to the cells they infect, they are nevertheless only the bearers of this information. Notwithstanding the claims to the contrary of a certain informational view of life, viruses are parasites, which, because they have no metabolic activity of their own, cannot properly be considered organisms.

A change of time and scale invites us to inquire into the relation between a complex multicellular organism and the cells out of which it is formed. This is not a new problem: ever since the development of cell theory in the nineteenth century it has been accepted that individual cells surrender a part of their identity to the organism as a whole. Until recently the problem remained relatively abstract and inconsequential. The development of two lines of research, however, each with different origins and implications, has forced scientists to reconsider the relation between the organism and its constituent cells, and, more generally, the familiar division of life into two levels or realms, the microscopic and the macroscopic.

The first line of research relates directly to human beings. The moment of a person's death was long thought to coincide with the death of the body's cells. A person was declared dead when the heart stopped beating—that is, at a point when most of the body's cells were either already dead or were condemned to die in the minutes that immediately followed owing to a lack of oxygen and nutrition. All that changed once it became possible to artificially maintain respiration and blood circulation in a body that was no longer capable of either one, and with the development of organ transplantation, which has a far greater chance of success if the organ is removed from a body that has been kept alive by artificial means. New legal and medical definitions of death were therefore needed. Current definitions are based on a number of tests that demonstrate the irreversible cessation of cerebral function, with the result that today a person can be declared dead even though most of his or her cells are still fully alive.

Our understanding of the relation between the death of the organism and the death of its cells has been further transformed by a second line of research, with corresponding implications for the relation between cellular life and the life of the organism. Beginning in the mid-1960s, scientists discovered a phenomenon known as programmed cell death (more formally, apoptosis—although, strictly speaking, this term denotes only one of the ways in which programmed cell death occurs). During the course of an organism's development, both before and after birth (or eclosion), a large number of cells will die.[13] This death is not accidental but rather, as its name implies, deliberate, enabling the organism to tailor its structures, or to adapt the number of its cells to its physiological state. The same process is involved both in optimizing neural connectivity in the nervous system and in controlling the activity of the immune system.

Much of the early observational evidence regarding programmed cell death is more than a hundred years old, but for a long while it was neglected. In the nineteenth century it was virtually inconceivable that death could be a normal part of the life of an organism, an indispensable component of embryonic development. Even in its smallest forms, death seemed obviously—almost by definition—to be confined to the end of life.[14] The evolution of multicellular organisms can now be seen, however, to have introduced a top-down system of control in which the fate of the organism dictates the fate of individual cells, rather than the other way around: it is the vitality of the organism as a whole that determines whether its constituent parts are alive or dead. Life, by its very nature, is a holistic phenomenon.

It has often been argued that, in order to have a clear idea of what a living organism is, one should focus on monocellular organisms, particularly bacteria. The problem is that the life of a bacterium is far from being fully autonomous. Many bacteria can form biofilms that prevent the healing of infected wounds, for example, in which the bacteria acquire new properties.[15] Similarly, many studies have demonstrated the central role of mitochondria in the phenomenon of apoptosis. One possible explanation is that cell death existed in bacteria, and invaded the precursors of eukaryotic cells during the period of endosymbiosis that gave birth to the modern eukaryotic cell.[16]

The meaning of programmed cell death for an isolated bacterial cell is not clear, however. There are several mechanisms of cell death in bacteria. For example, the fact that the loss of a plasmid (a small parasitic molecule of DNA) can automatically lead to the death of a bacterial cell suggests that programmed cell death may have been an evolutionary invention designed to preserve the presence of such DNA molecules in the host population. But this is probably not the only explanation of an event from which individual monocellular organisms obtain no benefit. Other mechanisms of cell death, coded for by the bacterial chromosome independently of the presence of a plasmid, are under the control of recently discovered factors that enable a bacterium to "measure" the number of members of the colony to which it belongs.[17] Programmed cell death will occur, for example, when the supply of nourishment is suddenly reduced, allowing a community to survive by limiting its number of individuals and providing the surviving bacteria with a valuable food source—the remains of dead cells. These mechanisms make sense only if the benefit they confer upon a bacterial population as a whole is taken into account.

In order, then, to understand the behavior of strictly monocellular organisms (which include most bacteria), we have to place them in the context of a living community that controls their fate. The ecological view of the world as a totality that functions in accordance with its own rules, rather than as a mere sum of individuals, casts a new light on the appearance of terrestrial life: it is no longer a question of which conditions permitted the formation of the first cell, but instead of which conditions permitted the appearance of the first living ecosystems. Note, too, that by adopting this perspective it becomes possible to pose the question of life on other planets in a new and different way.

It will perhaps now be clear why I have chosen to discuss an apparently disconnected group of topics in this chapter.

Taken together, they show that the problem of life cannot be considered solely with reference to the properties of an autonomous and isolated organism. Nor are the interactions between organisms merely the product of preexisting life forms; they are part of the very possibility that these organisms should be alive at all. This dialectical relation between autonomy and totality is therefore a key characteristic of life, for life was a system from the moment of its inception, and it has remained a system in its most evolved forms.

The formation of autonomous beings was an important, but nonetheless secondary, step in the process that allowed life to appear; what was essential was the appearance of a self-organizing system. Eventually this system produced individual organisms that then established between themselves the same kind of relations that had existed between the components of the original autopoietic system and their environment. This led in turn to symbiotic arrangements, lateral exchanges of material and information, and a hierarchical form of organization between autonomous cells. The systemic character of the relations between organisms is a reflection of the systemic character of life itself.

Part IV
A Few Necessarily Provisional Conclusions

In the second part of the book I briefly reviewed the main answers, ancient and modern, that have been given to the question "What is life?" The lack of any consensus was plain. The hopes raised in the middle of the twentieth century by the informational conception of life—which led some biologists to suppose that the question had disappeared—have now faded, and the old problems remain with us today, or so it would seem.

I then went on, in the third part, to consider some of the assumptions that guide current research into the origin of life on Earth as well as the search for extraterrestrial life. Here the absence of any serious reflection about what is actually being looked for is sometimes cruelly obvious. Some of the work now being done is based on one of two false assumptions. The first is that the best way to discover life's constituent principles is to examine extreme conditions, which fix the limits beyond which life is impossible. The organisms living in these conditions are often atypical, however, and in any case are the result of a long arc of adaptive evolution. The second illusion is that life can best be understood simply by studying isolated individuals. In general, however, the characteristics of living organisms can be explained only by taking into account the place they occupy in the various ecosystems of which they are a part.

On the other hand, the three features that most students of the problem agree are fundamental—the ability to reproduce, the existence of complex macromolecular structures, and the presence of an active metabolism that can synthesize these macromolecules—remain firmly established, even if, for experimental reasons, the study of metabolism has enjoyed the most attention and forms the basis for most contemporary astrobiological research. Before examining these characteristics more closely, however, I shall have to directly address the view, which until now I have dealt with only in passing and which threatens to undermine the very premise of my book, that the search for an answer to the question "What is life?" is not a rational and scientific enterprise.

13
Objections and Replies

A number of objections can be brought against the attempt to inquire into the nature of life. Some readers may be inclined to question the very existence of the phenomenon of life itself. Everyone will concede that in nature there are objects that are called organisms. Most of us would agree on the set of objects that can be described in this way, which by itself is enough to warrant scientific interest in them. Beyond that, however, it might be argued that trying to grasp precisely what these objects have in common is an idle, pointless, and ultimately nonsensical exercise.

This kind of criticism recalls the ancient philosophical dispute, particularly lively during the Middle Ages, over the question of universals: do the categories that we use to refer to the objects around us ("bed," "table," "tree," and so on) have any reality of their own, apart from the objects themselves? In the present case, the question is whether the general concept signified by the term "organism" exists independently of the various physical instances it denotes. If the term is merely a conventional device—a shorthand way of designating a set of objects, the composition of which most everyone agrees upon—then the concept "organism" has no real existence, no ontological reality. In that case, as philosophers say, it is not a natural kind or category.

The nature of the problem will become clearer if we examine the related concept of species. Do animal and plant species exist, apart from the individual animals and plants picked out by the term "species"? Or is the use of this term merely a way of ordering the diversity of the living world? The Darwinian revolution, even though it breathed new life into this debate, failed to resolve it, and biologists continue today to argue about whether species actually exist or not.[1] Similarly, the dispute over universals is an example of a "saturated" problem: all possible positions have been staked out and endlessly defended against attack, without producing any general agreement.[2] Since philosophers have yet to agree among themselves, it seems to me that we can safely leave this question to one side and carry on with our own investigation.

The question reappears, however, in a slightly different form in poststructuralist criticism. Michel Foucault had argued, a few years before Georges Canguilhem, that life did not exist before the birth of biology in the early nineteenth century.[3] For Foucault and others, the question "What is life?" is an artificial (rather than a natural) question, because it was formulated in the first place by human beings—a position that clearly derives from the constructivist conception of science (and, more specifically, from the relativist branch of this school that, sadly in my view, often speaks on its behalf).[4]

At a first approximation, the modern (or postmodern) version of the first objection, concerning universals, is easily answered by pointing out that life as a recognizable phenomenon certainly did exist before biology. To go further than this, however, one needs first to accept that scientific knowledge is in fact a human construction. In one sense, of course, it obviously is: as we have seen, human societies have asked (or not asked) "What is life?" in various ways throughout recorded history. The mistake is to conclude from this fact that the question itself is therefore of little intrinsic interest. Only those who seek an absolute form of knowledge are primarily concerned with the historical aspect of a scientific question. Scientists, who realize that both the questions they ask and the answers they give are necessarily provisional, do not normally insist upon it.

There is no reason to doubt that life is a precise scientific concept, or that research on the nature of life deals, however inadequately and incompletely, with real, actually existing objects with specific properties.[5] To claim this much amounts to embracing not so much philosophical idealism as the pragmatic tradition of natural science. For those who hold that life is not a natural category do not have more and better arguments at their disposal than those who defend the opposite position. One has only to consider the long history of scientific investigation down the centuries, which has produced ample proofs of the independent existence of the phenomenon of life. At this point, then, the burden of proof must fall upon those who assert the contrary.

A second objection to the basic premise of this book has a long scientific pedigree. In recent times it has been particularly well stated by François Jacob, who pointed out that modern science was born with the decision to abandon inquiry into general propositions in favor of the study of precise, simple, and limited questions.[6] Forgoing the temptation to ask what makes life unique is precisely what has made possible the explosive growth of the life sciences, Jacob argued, in the same way that refraining from speculation about the essence of human nature has permitted the development of the humanities in its many branches. Even if scientists decline to consider the most general questions, however, these questions are not therefore worthless—quite the contrary. As Darwin put it: "It is no valid objection that science as yet throws no light on the far higher problem of the essence or origin of life. Who can explain what is the essence of the attraction of gravity? No one now objects to following out the results consequent on this unknown element of attraction."[7] It was an unknown element—but undoubtedly a real one that, in the event, well rewarded further study.

Darwin's comparison of investigations into the nature and origin of life with the study of gravity is doubly interesting. On the one hand, as Darwin recognized, general questions are only grudgingly given up: thus Newton, for example, was reproached by Leibniz for having introduced "occult qualities" into science. On the other hand, inquiry into fundamental phenomena is abandoned usually for tactical reasons, and then only temporarily. Not long after Darwin's time, as part of the general theory of relativity, Einstein suggested that gravity is a consequence of the structure of space-time. This turned out to be an extremely fruitful conjecture for physics. In much the same way, the question of life, while it is not a good question for doctoral candidates in biology to explore, is a very good question for their teachers to examine.

There is a third reason for taking issue with my approach to the matter. Many of those who have shown the greatest interest in the question of life during the last two hundred years have wanted to draw a clear distinction between the animate and inanimate worlds, so that the former constitutes a separate domain outside the scope of the physical and chemical sciences. In reaction to this view, by insisting that all organic phenomena are material phenomena, scientists were able in their turn to exclude appeals to "vital" or "spiritual" forces from the logic of biological explanation. This is the main reason, as we saw earlier, why the question "What is life?" fell out of circulation for much of the twentieth century. The fallacy here lies in supposing that because organic phenomena can be studied scientifically, they do not have special properties of their own that need to be described as precisely as possible. To deny that life is a proper subject for scientific investigation risks throwing out the baby with the bathwater.

A fourth and final objection, related to the last one, proceeds from the assumption that scientists should be interested only in questions that can be answered experimentally. In order to answer the question "What is life?" one would therefore have to develop an experimental procedure that makes it possible to distinguish between life and non-life, with an organism being defined as an object that gives a positive response to such a procedure. This view is scientifically impeccable, but it conceals one crippling weakness: to develop experimental protocols in the first place, and then later to interpret their results, scientists must have some prior conception of what life is.

In further reply to these various objections, and in a more positive vein, I would say that there are at least two good reasons to wish to address the question "What is life?" The first is the very existence of a science of life—biology. Surely it must be considered paradoxical that so many people should devote their professional lives to the study of organisms without ever explicitly asking about the specific nature of the objects with which their research is concerned. In the short term, the lack of curiosity displayed by most biologists poses no problem: one can, after all, do some very good biology without wondering about the nature of life. In the long term, however, this indifference becomes a worrying sign of intellectual impoverishment, which ultimately will have a harmful effect on scientific creativity. To one degree or another, the educational system in virtually all nations has failed to give science students the background they need to think intelligently about fundamental—and fundamentally philosophical—questions of science, despite the years they are obliged to spend in lecture halls, seminar rooms, and libraries. This narrow conception of the nature of science is certainly one of the reasons why fewer and fewer students in developed countries today are electing to study it. The result is that such questions are left to others, who very often have no scientific training. Books by pseudo-scientists on the mysteries and wonders of life enjoy a flourishing popularity. Indiscriminately mixing fact and fantasy, these authors give a quite false impression of what science is, and what it is that scientists do.

A second reason for wanting to engage in a serious and conscientious way with the question of life is found in the new avenues of research opened up in recent years by astrobiology. If we are to properly assess the value of present and future discoveries, which at first sight are likely to be highly ambiguous, devising an adequate definition of life will be essential. This does not mean simply providing a list of all the characteristics that are associated with life. Many writers, as we saw in the second part of the book, have done little more than assemble a catalogue of the properties of organisms. It is not enough, as most scientists will admit, to describe; one must also choose, from among all the relevant properties, the one (or ones) whose possession entitles us to say of a particular object that it is a member of the living world.

In the chapters that follow I want to consider a question that follows naturally from what has gone before, whether the lack of a straightforward, widely accepted answer to the question "What is life?" is a sign of weakness from the point of view of biological theory. Regarding the origin of life, we have noted the application of two general strategies: the first, due to Aleksandr Oparin, extends the power of natural selection to the prebiotic world; the second appeals to some principle of self-organization and increase in complexity in nature. It may be, then, that we fail to recognize that the correct answer—Darwinism, many would say—is already known to us, or that it has only recently been intuited. There can be no question that complexity is one of the main characteristics of organisms. Perhaps the new and rapidly growing field devoted to the study of this phenomenon holds the key we are searching for.

For the moment at least I remain skeptical that either Darwinism or complexity theory will provide a satisfactory answer to the question of life. This is not a cause for despair, however. The outlines of an answer can already be discerned, I believe, in the form of three cardinal properties—metabolism, complex molecular structure, and reproduction—that we have already discussed at length. But we must be careful not to try to put these three characteristics into any kind of order, making one of them fundamental and the others secondary, in the manner of the ultra-Darwinians. We need also to keep in mind that life is a process, and that it is therefore perpetually being transformed. Life on Earth today is not the same as life a billion years ago, which in turn was different from life in its earliest phase three and a half billion years ago. The characteristics of organisms that we observe today cannot be assumed to be essential features of a minimal description of life, even if they are the result of organic evolution.

14

Darwinism in Its Proper Place

No scientist seriously denies Darwinism's explanatory power with regard to evolutionary phenomena. Indeed, it is the only theory capable of explaining the development of organisms on Earth. But this does not mean that Darwinism by itself answers the question "What is life?" To claim otherwise is to confound a well-established scientific theory with an ideology. My purpose in this chapter is to try to clearly distinguish the two things, and to examine the cultural and social reasons why it is so easy to move, often without realizing it, from the scientific theory of Darwinism to the ideology of what the American paleontologist Niles Eldredge has called ultra-Darwinism.[1]

When I speak of "Darwinism," I am not referring to the theory set out by Charles Darwin in *The Origin of Species.* Darwin's Darwinism was open to the idea of the inheritance of acquired characteristics, and natural selection was conceived, somewhat naively it must be said, as a struggle for survival between organisms. The Darwinism that I have in mind is the "modern evolutionary synthesis" that developed over the course of three decades, from 1930 or so through the 1950s, as a result of the convergence of work in fundamental genetics, laboratory and field population genetics, and paleontological studies of the transformation of life forms throughout the Earth's history. The American biologist Richard Lewontin has summed up the modern synthesis by reference to three principles:

1. Different individuals in a population have different morphologies, physiologies, and behavior (phenotypic variation).
2. Different phenotypes have differential rates of survival and reproduction in different environments (differential adaptation).
3. There is a correlation between parents and offspring in respect of their contribution to future generations (heritable adaptation).[2]

This is a coherent theory that explains the vast majority of adaptive phenomena found in the living world, unlike the Lamarckian theory of inheritance, according to which modifications acquired during the life of an organism can be transmitted to its offspring. Darwinism in the form Darwin gave it said nothing about the limits within which morphology, physiology, and behavior may vary. It did not, for example, exclude the possibility that organisms are constrained to evolve within a range of possible forms. Nor did it exclude the possibility that contingent events play a decisive role in the evolution of life. The genetic makeup of a population, for example, may undergo spontaneous, undirected change in the absence of any selection pressure, through genetic "drift." Similarly, the genetic characteristics of a population, and its evolution, will depend heavily on the nature of the organisms originally found in it if they were relatively few in number. All such phenomena are fully integrated into the evolutionary synthesis.

Over the past four decades a great many studies and models have challenged various aspects of this synthesis. Epigenetic studies have described reversible modifications of DNA and chromatin affecting gene expression that can be transmitted, at least in part, to later generations.[3] The importance of phenotypic plasticity has now been very carefully established, building on the fruitful intuitions of the Russian evolutionary biologist Ivan Schmalhausen, the American psychologist James Baldwin, and the British embryologist and geneticist Conrad Waddington. An initial phenotypic change is considered the first major stage of adaptation, preceding the stable variation of the genotype that will finally fix a new phenotype.[4] The environment can alter the rate of mutation, but not reduce its randomness. Conversely, organisms can alter their environment by creating "niches," the characteristics of which are transmitted to succeeding generations.[5]

Taken together, these new perspectives have generated a far more dynamic picture of the relations between organisms and their environment than the standard Darwinist account.[6] But do they really undermine this account? Do they represent the first step toward a new synthesis? This might turn out to be the case if a general law of complexification were to be discovered, operating throughout the universe and in particular with regard to organisms. But most scientists doubt the existence of such a law. For the moment, at least, the result of recent work seems to have been to revise and enrich the modern evolutionary synthesis, not to shake its foundations.

Two other phenomena that I discussed earlier also appear to run counter to this synthesis.[7] The first is the horizontal transfer of genes, as against the Darwinian view of the transmission of characteristics by descent. There is evidence, as we have seen, that genetic innovation may not be a product of natural selection, instead being rapidly shared among organisms by a Lamarckian process.[8] This does not constitute a serious problem, however, at least as far as "higher" organisms are concerned. Human beings, for example, receive virtually all their heritable characteristics from their parents. The situation is not quite so clear-cut in the case of microorganisms: the rapid spread of resistance to antibiotics, which is largely due to the horizontal transfer of genes through the action of small DNA molecules (plasmids), shows the importance of non-Darwinian processes. Matters are even less straightforward when one considers the earliest stage of life, where horizontal gene transfer may also have played a major role. At this stage of evolution it is more exact to speak of a gene pool rather than of genetic units. It is possible that viruses, which, as we have seen, circulated with ease among protoorganisms, were crucial to the exchange of genetic information in the first living world. Subsequent innovations, such as the appearance of DNA, may have been due to viral activity.

The second phenomenon that does not neatly fit into the Darwinian schema is symbiosis. Mitochondria and chloroplasts, for example, are the product of symbiotic relationships. Other symbiotic phenomena may have led to the appearance of other parts of the eukaryotic cell. Furthermore, the isolation of the first living systems was probably not absolute, and it is likely that precise mechanisms of reproduction did not yet exist. It seems reasonable to suppose that the evolution of these early cells took place more often through fusion and fission than through a precise process of division. If so, the evolutionary synthesis loses its explanatory power when applied to the earliest stages of life.

One of the outstanding characteristics of Darwinian theory in recent years is its tendency to invade other domains of knowledge. One thinks, for example, of the attempt to extend the model of variation and selection to molecular and cellular activity. The French biologists Jean-Jacques Kupiec and Pierre Sonigo have urged, for example, that the worn-out notion of a genetic program be replaced by a series of models in which random processes intervene at each stage of embryonic development and cellular differentiation, generating different molecular or cellular states that are then selected during development.[9] These models have the virtue of limiting the number of instructions required to generate a new differentiation state. Additionally, they help to explain some puzzling observations that have been made regarding various processes of differentiation. But it is difficult to see why anyone should suppose, on the basis of a few sparse and preliminary results, that they are capable of explaining the process of cellular differentiation in its entirety. Nor is it clear why promising molecular approaches (in particular, the search for gene networks that regulate development) should be rejected out of hand, and still less why the rise of molecular biology over the past few decades should be condemned as a step backward. While one must be prepared to admit the existence of such random mechanisms, it is far from obvious why it should be necessary, as these authors insist, to prefer them in all instances to genetic mechanisms, renouncing any attempt to devise a deterministic model of differentiation.

At the molecular level, temperature-based agitation is seen as providing a range of states from which organic macromolecules choose those that are most favorable to a given kind of molecular activity. These random variations, which are nothing more than the incessant movement of molecules known as Brownian motion, are held to explain everything from enzymatic activity to the action of molecular motors and the passage of molecules and macromolecules through membranes.[10] Nonetheless, they fail to account for the molecular complexity of organic phenomena. It is quite true that random molecular variations occur, and that organic molecular systems have to cope with them, as it were, either by taking advantage of them or, more often, by working against them. They have no choice. But in comparing Brownian motion to the genetic variation of organisms, and the stabilization of certain of its states to natural selection, the meaning of the terms "variation" and "natural selection" has been significantly altered. This hegemonic tendency is a sign that ultra-Darwinism has become more of a fundamentalist ideology than a genuinely scientific theory.[11]

All Darwinian approaches, even ones applied to topics of research far removed from the evolution of organisms, have in common the idea that order can appear through the selection of random variations. The concepts of variation and selection were formulated by Darwin himself in a very casual way, although it should be said in his defense that the true nature of variation only became apparent some decades later with the rise of genetics. This lack of precision led some early opponents of Darwinism to accuse it of substituting tautologies for explanations, and indeed a more rigorous definition of these central concepts required many years of work.[12] The modern ideological version is based on two assumptions: first, that there are a nearly infinite number of states in nature; second, that an efficient process of natural selection operates upon these states. Even if the space of possibilities that is supposed to be open to variation is not quite infinite (contrary to what many of its adherents seem to assume), it is in any case so large that any constraints which might exist are deprived of virtually all their effect. There is something almost Nietzschean about the power assigned to variation in the ultra-Darwinian vision of life, which magnifies the role of chance—what is sometimes called the "inventive" power of nature—to the point that there is no room left for any ordering principle to come into play.[13]

The notion of selection, for its part, is often reduced to the idea that some variants are better than others in the sense that they make it possible to explain how a particular process led to the structure or function we know today. Sometimes it is used simply to mean that some variants replicate more rapidly than others. In that case one should follow the late biochemist Leslie Orgel in speaking of "functionless natural selection"—that is, selection having no significant function apart from the power of self-replication.[14] But it is not clear that one can legitimately use the term "Darwinian" to describe a process of selfish multiplication of the sort displayed today, for instance, by certain DNA fragments in the genome. To suggest that competition between two organisms in a given environment is similar to that between two molecules with different rates of self-replication comes very close to saying that selection with function and selection without function are the same thing. My own view is that the two processes are related, but nonetheless distinct. We have seen that, depending on which scenario of the origin of life is adopted, selection without function—that is, the appearance of selfish replicators bearing no other functional capacity than that of self-replication—either preceded or followed the appearance of selection with function. As in the case of viruses, which cannot have been responsible for the origin of life, even if they may have played an important role in its development, it seems more plausible to assume that the appearance of selfish replicators is a by-product of life, a kind of pathological fruit of its reproductive power, rather than a stage in its formation.[15]

In addition to these "external" forms of selection, ultra-Darwinians sometimes refer to "internal" selection. By this it is meant that not all variants and all combinations are possible; some are incompatible with the existence of a given system. To speak of the internal selection of chemical compounds, for example, is a way of saying that the chemical combination of elements obeys certain rules.[16] Paradoxically, the existence of regularities and constraints now becomes part of the power of natural selection, not something that limits its power. Here again, the use of the term "selection," which goes well beyond anything Darwin actually said, makes it clear that ultra-Darwinism is not a scientific theory but an ideology attached to a highly specific ontology.[17] It is, of course, quite true that extending the use of a concept beyond its customary field of application has frequently been a major impetus in the development of science. Nonetheless, the fact that such a departure is being made must always be borne in mind, and great caution exercised in applying the concept to its new domain. Nowhere is this more true than in adapting the concept of selection to the prebiotic world.

There are a number of reasons why ultra-Darwinism currently enjoys such favor. Ever since its original statement in 1859, and despite the various transformations it has undergone in the interval, Darwinian theory has always been closely associated with the values of the societies in which it has been developed and taught. In Darwin's own time, the doctrine of social Darwinism was used to justify the structure of Victorian society and the dominant position of European civilization in the world. Today, the view of organisms as self-interested opportunists corresponds to the free-market model of social success based on the pursuit of individual economic advantage. What might be called the "socially correct" idea that diversity is a source of wealth (the wealth of nations, after all, is conventionally supposed to derive from the economic diversity of countries and individuals engaged in trade with one another) usefully conceals the increasing standardization of behaviors and habits under the influence of economic globalization.

The "natural" sympathy that many scientists feel for ultra-Darwinism is reminiscent of the reaction of children taken to the zoo for the first time: confronted with "the bewildering variety of the animal kingdom," in the words of Jorge Luis Borges, they find it a cause for joyous delight rather than for fright or horror. A similar sense of pleasure in the face of the exuberant power of life seems to dispose grownups to accept ultra-Darwinian explanations, though this may also be due to the relative poverty of adult imaginations—a point that Borges also emphasizes ("the zoology of dreams is far poorer than the zoology of the Maker").[18] The vogue for ultra-Darwinism is a consequence, too, of the apparent successes that it has enjoyed in certain fields of research. Applying the concepts of variation and selection to molecular activity has produced a new branch of chemistry called combinatorial chemistry, which in the early 1990s inspired a great deal of research into new medicines. Combinatorial chemistry involves randomly varying the structure of molecules to generate all possible combinations of substituents and then analyzing the compounds synthesized in this way, in the hope of being able to detect the very few that exhibit interesting pharmacological activity. Although the chance that a particular molecule will display such activity is exceedingly small, this disadvantage is offset by the very large number of molecules that can now conveniently be tested. By comparison with conventional laboratory methods, it is far more efficient to rapidly explore a virtual space of possibilities than to draw up step-by-step maps.

Combinatorial methods are very useful in analyzing nucleic acid polymers (especially RNA molecules), where a large number of different chains can be synthesized despite the fact that the space of possibilities consists only of four solutions at each position in a given chain. The selex method, as it is called (a shorthand for "systematic evolution of ligands by exponential enrichment" or, more simply, in-vitro evolution), has made it possible not only to isolate RNAs with new catalytic or fixation capacities—thus demonstrating the "creative" power of variation and selection—but also, at least in the minds of ultra-Darwinians, to validate the hypothesis of an RNA world.[19] The mechanisms of selection exploited by this sort of technology are highly artificial, however. To conclude from experiments performed using the selex method that natural processes of variation and selection have comparably great creative power is to make an even bigger leap than Darwin did in likening natural selection to the kind of artificial selection practiced by animal breeders of his time in Great Britain. The quality and power of the artificial selection mechanisms that have recently been developed, as well as the wealth of random variations they involve, are unquestionably very impressive, but it does not follow from this that nature imitates them.

The superficially successful extension of the modern Darwinian synthesis has encouraged hopes in recent years of formulating a general theory of the evolution of the physical world, in both its organic and inorganic domains, by variation and selection. A similar expectation had been aroused in the nineteenth century with the development of thermodynamics. The seemingly natural longing on the part of a great many researchers today for a single explanatory theory—a "theory of everything"—has not dimmed despite the growing fragmentation of scientific knowledge. It is nonetheless, in my view, an illusion. The notion that a general theory of the physical world can be based on ideas as vague and imprecise as the ones advanced by ultra-Darwinians is implausible enough, even foolhardy; and the application of the mechanisms of variation and selection to the realm of ideas and theories, attempted by a number of biologists (among them Jacques Monod and, most notably, Richard Dawkins), has shown itself to be more doubtful still.[20] Indeed, the very fact that ultra-Darwinism can be applied to both living and non-living phenomena is proof that it cannot claim to supply a sufficient answer to the question "What is life?" for the criterion by which life is to be distinguished from non-life, among all the objects subject to variation and selection, remains undefined—thus paradoxically once again undermining the ambitions of this doctrine within the field of biology itself.

In holding that the chemical nature of organisms and their metabolic characteristics are secondary by comparison with the power of reproduction, Darwinism in its extreme version actually makes it impossible to draw a clear boundary between life and non-life. Nor can ultra-Darwinism deny artificial organisms membership in the kingdom of life. It may be objected, of course, that the "beings" which can now be created in a computer are artificial only to the extent that the study of such creatures is still in its infancy, and that after a sufficiently long period of evolution they will be no less complex and vital than the organisms of the natural world.[21] Most biologists, however, believe that artificial life, no matter how great its value may be in mimicking certain characteristics of organisms, and thus making it possible to study those characteristics, is nevertheless itself not truly alive. Similarly, it is difficult to exclude viruses from the living world on ultra-Darwinian assumptions.

Ultra-Darwinians always make the same mistake: they ignore constraints—in this case, the chemical constraints that are the very basis of organic phenomena. Classical Darwinism elucidated what might be called the rules of the game of life, namely, stochastic variation coupled with selection of those organisms that most successfully reproduce themselves in a given environment, a cycle that in turn promotes adaptation to this environment. But ultra-Darwinians act as though the emergence of these rules, and the circumstances under which they became possible, are not part of the explanation of life itself. The question "What is life?" forces us to take into account both the nature of the basic elements of life and the chemical transformations that are necessary for the synthesis of these elements. Life, whatever else it may be, is not a disembodied phenomenon. Nor is the ultra-Darwinian answer to the question *the* answer.

15

The Lure of Complexity

The variety of answers that have recently been given to the question "What is life?" is perhaps the clearest sign that a genuine theory of life is still wanted. That, of course, is precisely what Schrödinger tried to develop more than a half century ago—a physical theory of life based on molecular order and negentropy.

Twelve years before the appearance of Schrödinger's book, in 1932, another founder of quantum mechanics, the Danish physicist Niels Bohr, delivered a seminal lecture entitled "Light and Life."[1] Like Schrödinger, Bohr sought to exploit the new conceptual arsenal of physics to explain the phenomenon of life, but his solution differed from the one that Schrödinger later advanced in *What Is Life?* For Bohr, just as matter behaved at the elemental level both as a wave and as a particle, and thus had to be studied in these two complementary forms, so too the phenomenon of life had to be studied using both a reductionist, physicochemical approach and a more global approach that would capture its uniquely irreducible character. The first method had already been explored for many years; the second could be undertaken only if a new life science was created.

This was the task that a young physicist inspired by Bohr's ideas named Max Delbrück set himself when, in 1937, he began to study bacteriophage reproduction, which later played such an important part in the rise of molecular biology. Delbrück argued that life was characterized by its ability to reproduce. The simplest example of this phenomenon is bacteriophage replication, which therefore supplied a convenient model system for experimental research.[2] To Delbrück's great disappointment, intensive study of this phenomenon failed to reveal a new physical principle underlying organic life. It seems clear in retrospect that, like the ultra-Darwinians who came after him, his error was to identify life with a single property—the ability to reproduce—while ignoring the structural and metabolic characteristics of organisms.

The notion that life can be explained by a particular physical characteristic goes back much further than Delbrück. In the second half of the nineteenth century, isotonic force, surface tension, and even radioactivity were all proposed as explanations for life's distinctive features. In every case the same problem arose: these forces are also found under circumstances where life is quite clearly absent. Still today many physicists think that the profusion of molecular descriptions of organisms demonstrates the lack of an adequate physical theory. The phenomenon of life, they believe, must have a simple explanation. But they no longer seek this explanation at the microscopic level; instead they look to account for it at the global level in terms of complexity, systemicity, and self-organization.[3] In its most extreme form, this approach regards life as a critical moment in the general movement of the universe toward complexity, which becomes visible in the gradual formation of atoms from quarks, molecules from atoms, planets from clouds of matter, and so on. With the appearance of life, organisms came to embody and reinforce the same tendency toward increasing complexity.[4]

Evidence of the enthusiasm with which almost all scientists in recent years have embraced the idea that biology must go "beyond reductionism" may be found in the publication of thick supplements to the regular issues of the main scientific journals and the creation of institutes devoted to the study of complex systems at the most prestigious American universities.[5] As with all such pledges of allegiance, the current passion for complexity and its cousins springs from conflicting motivations that have nonetheless converged to form a heterogeneous coalition of interests. For some biologists, the study of complex systems is simply another step forward in the development of their discipline: characterizing genomes (that is, the totality of genes and their products) leads naturally to a search for ways in which elementary components interact structurally and functionally to endow organisms with their distinctive characteristics; post-genomic techniques, which seek to describe the simultaneous expression of all the genes in an organism, or the totality of protein-protein interactions that take place within it, likewise aim to provide a general account of how organisms function while substituting a dynamic picture for the present static one. For other researchers, the study of complexity is a way of taking revenge on the reductionism of molecular biology and promoting the return of a holistic biology that acknowledges the existence of distinct levels of organization in the living world. For still others, it represents the recognition of work begun over half a century ago, taking physics as its model, that sought to create a theoretical biology capable of providing a mathematical formalization of the processes that take place within organisms.

It is not surprising, then, that a number of very different, even discrepant approaches should be found under the banner of complexity theory: attempts to formalize biological processes that are already well understood at the physicochemical level (metabolic pathways, networks that regulate gene expression or signaling pathways, and so on) and then to recreate these processes with a view to manipulating them for therapeutic purposes, or else to synthesize "new" organisms designed to carry out specific tasks (what is called synthetic biology); comparative analysis of complex machines and organisms with the aim of discovering the sources of the "robustness" of organic life (its resistance to variations in the external environment or to the alteration of certain components); description of the geometric characteristics of social and biological networks; and agent-based modeling of social and economic behavior. This is by no means an exhaustive list.[6] In bringing together strands of research that have little in common, the study of complexity sometimes seems to mean anything new and incomprehensible to nonspecialists that comes from the world of physics!

Everyone will agree that organisms are both self-organized and complex. Kant long ago pointed out that each part of an organism exists only because of the others, for the others, and as a function of the totality they jointly compose. Recent advances in molecular biology have only disclosed a new—and, for some, disturbing—facet of this organic complexity. It is clear in any case that the structural and functional complexity of organisms is constitutive of life: life is situated *beyond* a certain degree of complexity. The question remains, however, whether the evident sophistication of organic development implies that a satisfactory explanation of the nature of life will finally be provided by some unifying theory of complexity, as Jack Cohen and Ian Stewart, in a review of recent advances in the search for intelligent extraterrestrial life, implicitly suggest in asserting that life is "a name we give to certain emergent processes of complex systems."[7] One difficulty immediately presents itself: what exactly is meant by "complexity"? The usual reply is that a system can be considered complex if its behaviors and its properties amount to something *more* than those of its component parts. In other words, something *emerges* from the integrated functioning of the system. What precisely is to be understood by "more" and "emerges" in this connection has been the subject of much debate, however.[8] "More," for example, might mean merely that because we know so little about the parts of the system, the behavior of the system as a whole seems surprising to us.

Among those who uphold this apparently new way of looking at the living world, two opposing attitudes may be detected.[9] Some feel that the reductionist approach to the study of organic life that until recently has dominated biological research must be considered a failure. The global approach is seen as something very different, turning away from a concentration on elementary components in order to discover general principles underlying the function of complex systems. For others, order emerges from local interactions and relations within a system that must first be understood in order to describe the function of the system as a whole.

The view of organisms as complex systems has the advantage of making it possible intuitively to grasp the origin of a number of life's most salient characteristics. The threshold of complexity corresponding to the formation of an autonomous organism, for example, requires that it have at least a certain minimal size and a certain range of molecular diversity. For the moment, however, complexity theories influence research in biology mainly on the level of metaphor. Thus, for example, the American scientist Stuart Kauffman, perhaps the leading theorist of biological self-organization, describes the appearance of life as a phase transition that occurs once a critical threshold of complexity has been reached.[10] Under his influence old words such as "integration" have become outmoded, while new words like "connectivity" have been invested with great authority—so far more rhetorical than explanatory.

And yet it is hard not to be fascinated by Kauffman's writings. His account of the origin of life, reconciling the genetic and metabolic scenarios, is one of the most elegant yet proposed. Kauffman holds that the first stages of metabolism involved macromolecules (rather than small molecules) and reactions in which each macromolecule contributed to the formation of one or more other macromolecules. His conviction that he has found *the* answer to the question of life is infectious: more than once I have found myself wondering whether I should really go on resisting his arguments, which have been impressively corroborated by a good many computer models. And yet once the impression of revelation induced by such phrases as "life is an autocatalytic system carrying out one or more work cycles" has dissipated (which takes some time, given Kauffman's great skill as a writer), one's doubts return.[11] Is this not simply another way of saying that organisms can synthesize their own components and perform various tasks by using energy from the immediate environment—something that biochemists have been saying for more than half a century?

Redefining a problem sometimes constitutes a major conceptual advance. In this case, however, redefinition seems to have obscured a clear view of life's specific characteristics and replaced it with a definition that can be applied to a whole range of systems, many of which have nothing living about them. As with the ultra-Darwinian argument, what appears at first sight to be a powerful, simple, and definitive explanation of the world turns out, once one tries to grasp it, to be nothing more than an elusive shadow, a phantom.

Past experience should perhaps make us somewhat skeptical about the prospects for applying theories of complexity to organisms in the years ahead. The idea that organisms are complex systems is hardly new; nor is the appeal to such theories to rescue biology from the dead end into which reductionism is said to have dragged it—indeed, the Austrian-born theoretical biologist Ludwig von Bertalanffy's general system theory is almost a half century old by now.[12] An example will clarify what I mean. Organisms constitute special systems from the point of view of thermodynamics, because they continually exchange matter and energy with their environment; in the technical phrase, they are a kind of open system far from thermodynamic equilibrium. After the Second World War, the Belgian chemist Ilya Prigogine showed that such systems could produce a form of organization that he called "dissipative structures." One thinks, for example, of Bénard instabilities, named after the French physicist who demonstrated that such structures appear within a heated liquid (where temperature differentials create upward and downward convection currents exhibiting regular geometric patterns of flow).[13] Although thermodynamic disequilibrium can order matter through a process of self-organization, the resulting structure disappears as soon as equilibrium is reestablished. The striking thing about structures created in this way is that they can randomly assume any one of many states.

The thermodynamic disequilibrium that characterizes organisms is undoubtedly part of the reason they are able to develop complex structures and functions. A small number of phenomena belonging to the living world, all of which lend themselves to this kind of interpretation, have been regularly put forward as examples of open far-from-equilibrium systems since the 1950s. But this argument is based on nothing more than similarity, and no convincing demonstration has ever been given that such phenomena can be explained by the thermodynamics of systems in disequilibrium. Indeed, the use of the term "self-organization" to describe both Bénard instabilities and the complex structures and functions of organisms may promise more in the way of illusion than explanation. At the very least it seems unlikely that the order of the living world, which has gradually emerged over the course of billions of years, can be reduced to transitory structures capable of appearing and disappearing in the space of a few seconds.

Even so, theories of complex systems are not without value—far from it. Some forms of complexity modeling, particularly ones involving networks of various kinds, will surely prove their usefulness in the interpretation and description of biological phenomena; already, in fact, they have been used to explain important phenomena in communications, economics, and social behavior.[14] But the theories and models of this new field of research, and the concepts upon which they are based, cannot be said to provide an ultimate explanation of the phenomenon of life—for the simple reason that an adequate explanation is already available to us.

16

The Three Pillars of Life

Having found the Darwinian answer to the question "What is life?" wanting, and equally little reason to expect that a decisive breakthrough will be forthcoming from theories of complexity, we are left to ponder the relatively broad consensus among scientists that organisms share three fundamental characteristics: they possess particularly complex molecular structures; they routinely support a large number of highly specific chemical reactions by drawing upon the external environment for both molecular material and energy; and they reproduce themselves.

These three characteristics are closely allied with one another. Complex molecular structures supply the catalytic basis for metabolic reactions and the ability to reproduce. Metabolism is responsible for the synthesis of macromolecules, which in turn catalyze its constituent chemical reactions. Inexact self-reproduction leads to an increasingly efficient adaptation of both molecular structures and metabolism to the environment. The first two characteristics thus provide what the late German-born evolutionary biologist Ernst Mayr called the "proximal causes" of biological phenomena; that is, their physico-chemical explanation.[1] The third characteristic, reproduction with variation (and its corollary, natural selection of variations), furnishes the ultimate cause of biological phenomena; that is, the explanation of organic development that has been forged by time and history.

A common error, as we have seen, is to try to rank these three fundamental characteristics of life, so that one is seen as primary and the others as being derived from it. The true relations between them are historical rather than hierarchical.[2] I have already criticized those who hold that reproduction is the foundation of the living world. Taking either structural or metabolic characteristics as the basis for a definition of life is equally without justification, however. It is, I admit, not altogether impossible that the self-organization of either complex metabolic systems or interacting macromolecules preceded reproduction. But the stability of life—the sine qua non of life as a historical phenomenon—is strictly and inalterably linked to its capacity for reproduction. This reproduction is inexact (which also implies that the possibility of life evolving is indirectly included in its definition), but nevertheless in general sufficiently faithful for the products of such reproduction to be functional.

Treating the structural and metabolic aspects as a single characteristic would amount to reviving the theoretical perspective of the early twentieth century, when terms like "protoplasm" and "colloids" were used to describe both the structural characteristics of cells and the functional (in particular, metabolic) properties associated with these characteristics.[3] This perspective had the virtue at least of putting an end to the rather unproductive dispute between those who, like Claude Bernard, considered the special character of life to reside in the particular properties of living matter, and those who, like Lamarck, sought it in the organization of matter. For the moment, however, it seems more sensible to regard molecular structure and metabolism as separate, albeit closely linked, pillars of life.

Distinguishing the two in this way acknowledges the fact that most proteins have functions other than metabolic ones, for instance in the organization of cell structure and the exchange of information and material within and between cells. Moreover, and still more importantly, this distinction recognizes both the difference between strong and weak chemical bonds and the central place of the latter in the living world (weak bonds are responsible for the process of self-assembly, which plays a major role in cell formation and has traditionally been opposed to self-organization in open systems).[4] But it would be a mistake to push these oppositions too far: there is a continuum between weak and strong chemical bonds, and between self-assembly and self-organization, no less than between the structural and metabolic aspects of organisms. Macromolecular structures and metabolic functions represent two poles of the chemistry of life.

The secret of life resides not in the precise nature of the chemical components of organisms, but in the systemic relationships that the chemical components of organisms jointly support. These relationships cannot be described in a general or abstract manner, however, as autopoietic theories assume, for they nonetheless depend on the very rich and specific details of the components themselves; indeed, the network of chemical relations that characterizes life on Earth could not exist without a certain type of molecule. The living world is therefore the product, both structurally and functionally, of a particular chemistry. The philosophical claim that matter structures reality is nowhere better supported than in the case of the chemical components of organisms, in which matter and form are completely intertwined.[5] Throughout evolutionary history, organisms have exploited the structuring power of weak chemical bonds, which arise from the specific characteristics of the basic elements of organic chemistry.

No scientist any longer seriously disputes the fact that life is an emergent property of systems, that organisms are systems, and that the structural and functional relations among organisms constitute systems. But this kind of description does not take us very far beyond the realm of metaphor, and replacing metaphors by abstract concepts will do nothing to help us better understand the nature of life. Quite the opposite: it serves only to hide the structural characteristics of the components of organic systems, which are essential if we are to understand the chemical properties of these systems. (It is for this reason, by the way, that it is a mistake to view the Earth as an organism, as the English scientist J. E. Lovelock does.[6] Earth is indeed a self-regulating system, formed by all the organisms found in it, its atmosphere and its oceans, and so on. Yet although the existence of this system must be taken into account in order to explain the evolution of the planet and of the organisms that it carries, the exceedingly heterogeneous nature of the elements of this system prevents it from being considered a true organism—this and the fact that it does not reproduce.) Answering the question "What is life?" may better perhaps be likened to a sculptor who tries to convey something of the essence of the sitter's personality by placing him in the company of the tools of his profession—a piano in the case of a composer, or, in the case of a painter, a palette and brushes. Just so, life cannot be summarized in a simple phrase or characterized by reference to a single property. All the elements of the natural world that are associated with life must be revealed if we are to understand it.

In order to exist, rather than merely subsist, life needs all three of the characteristics I have just mentioned, which are therefore essential to it. It can continue for a time in the absence of reproduction, even of metabolism; and structure, as we have also seen, may under certain circumstances be enough by itself to keep an organism alive. In this sense the three pillars of life are not equivalent. But the possibility that life can remain in a latent state through the conservation of structure alone is not an ancestral characteristic of organisms; it is the result of a long evolutionary process of structural complexification. Viruses, for example, are not alive, because they fully possess only one of the three necessary characteristics (structural complexity, together with a few of the elements required for reproduction). During an infectious phase, however, viruses are able to combine this single vital characteristic with the ability of their host cells to synthesize and reproduce, thus temporarily and incompletely acquiring the status of an organism.

Nor were the three pillars of life equally important at the beginning of life. In its earliest stages, as some suppose, life may have involved the formation of macromolecules, which in turn stimulated the synthesis of other similar or complementary macromolecules. Or it may be, as many others think, that life first arose with the appearance of self-stabilizing metabolic activity, which slowly evolved toward the formation of macromolecular structures, with the ability to reproduce following shortly thereafter. In either case, life would have emerged when all three characteristics had developed to the point that a set of relationships stable in both time and space came to be established among them. On this view, the isolated nuclei of life must have formed in an environment that was not itself alive, but that was already rich in complex molecular structures, chemical reactions, and replicatory processes, whether of macromolecules or vesicles. These sparks of life may have coalesced and then come apart, and sometimes gone out. But then others came forth again elsewhere, again only momentarily, until finally autonomous and stable organisms appeared.

It may be objected that my solution to the problem of life has a serious weakness, namely, that it considers life to have been born of a chance conjunction of distinct natural phenomena: the appearance of particular molecular structures, a series of intense chemical exchanges, and an autonomous capacity to reproduce. Yet this does not mean that life is a miracle: the phenomena involved, though very different from one another, are interrelated. The fact that life arises from the linking together of different orders of phenomena is the source of its distinctive character, and points to the role of an elusive random element. It is this condition of being linked together—not the allegedly awesome power of natural selection—that gives organic phenomena their special richness. I should hasten to add (although perhaps by now it goes without saying, since I have said as much many times before) that this view of the matter is inevitably provisional, and that it will continue to be qualified by further advances in scientific knowledge.

A few biologists, such as Thomas Cavalier-Smith at Oxford and Jack Szostak and Donald Ingber at Harvard, have already tried to clarify the way in which these three pillars of life came to be conjoined. Szostak seeks to establish a relationship between the properties of cell membranes involved in reproduction and the characteristics of metabolism.[7] Cavalier-

Smith and Ingber, though they reach different conclusions, both emphasize the role of the cell's internal structure, in particular its "skeleton," which they suspect served to bring together the membrane (and therefore cellular reproduction), the replication of macromolecules, and metabolism.[8] Such attempts quickly run up against an apparently insuperable obstacle, namely, our virtually total ignorance of the mechanisms that underlay the evolution of this cytoskeleton, and of its earliest forms. Inevitably, it is difficult to know whether the interrelationship of structure, metabolism, and reproduction at the cellular level observed today is ancestral, the modified remnant of something that happened at the beginning of life, or a later evolutionary invention that consolidated and reinforced a preexisting linkage that originally had assumed a different form. The appearance and the development of life can in any case be conceived only as processes, and therefore as historical phenomena. My answer, if it should turn out to be correct, will be correct only by virtue of the historical dynamic uniting these three pillars, which, as I say, remains to be elucidated.

An obvious way to demonstrate that we understand what life is would be to create life from elementary chemical components. Viruses have already been synthesized but, for reasons that I have rehearsed at length, this does not amount to synthesizing autonomous beings. Bottom-up approaches of various kinds have proliferated in recent years, all of them with the purpose of mimicking at least some of the properties of organisms.[9] One experimental approach to this problem might involve the creation of a vesicle formed of lipids in which suitable quantities of the some 200–250 types of protein whose functions have been shown to be essential for all known forms of life are combined, together with a DNA molecule coding for these proteins (including the appropriate regulatory signals), and which is then plunged into a chemical environment rich in elementary organic molecules. The project that Craig Venter has developed is but a first step in this direction, since it consists "merely" in injecting a synthetic chromosome into a preexisting bacterium that has been emptied of its own genetic material.[10] For the experiment to be decisive, the proteins encoded by the genome would, of course, have to be artificial; that is, they must bear no resemblance to any known protein, apart from their function.

This experiment is perfectly possible, at least in theory. As a practical matter, however, because the prospect of being able to efficiently synthesize artificial proteins with precise functions is still a long ways off, it will be hard to undertake any time soon. What is more, it will be necessary to synthesize not just one type of artificial protein, but more than two hundred, collectively displaying a wide range of functions. Finally, there is no guarantee that the cellular system created in this way would function properly. It seems probable that the first systems would work only for a short while, and that a great many adjustments would have to be made before the regulatory elements function as they should and the system becomes self-sustaining. These are the same sort of difficulties that physicists working on nuclear fusion encounter when they try to convert a reaction lasting a fraction of a second (which has been achieved) to a stable system similar to the one that exists in the inner core of stars.

For the moment, however, let us imagine that this kind of experiment were to be approved by the relevant ethics committees, and that it succeeded in producing a cell that was able to survive, even to reproduce.[11] This would clearly constitute a definitive demonstration that the phenomenon of life is no longer a mystery. The new organism, in its components and its functional principles, would nonetheless be a copy of organisms that exist on Earth. In other words, its creation would tell us nothing about the historical dimension of life, nor about its generality.

Conclusion

In adopting a triadic conception of life, I have left to one side a number of characteristics that figure in the various accounts that were presented in the second part of this book or elsewhere in passing: the hierarchical organization of the living world; its richness and diversity; the morphogenetic inventiveness of organisms; their robustness in response to extreme variations in their internal and external environment; the apparent purposefulness of their structures and behaviors; and even the fact that they can fall ill.[1] All these characteristics of life are quite obvious in complex organisms, although they are also present in simpler life forms.

The decision to ignore these characteristics runs counter to a venerable philosophical and scientific tradition extending from Aristotle, for whom higher organisms were the most perfect, up through Claude Bernard, who argued that the distinguishing features of life are much more pronounced in higher organisms. Two French philosophers of science, Dominique Lecourt and André Pichot, carry on this tradition today in their different ways. Lecourt holds that solid foundations have now been found for the theory of life in the molecular characteristics of the genes that control embryonic development.[2] Pichot, on the other hand, follows the thermodynamic tradition in maintaining that the special character of life lies in the increasing divergence between the evolution of organisms and that of their environment.[3] In both cases, the products of organic history are used to define life itself.

My own opinion is that these characteristics are emergent properties of organisms that appeared as a consequence of a long process of complexification. Even the remarkable evolutionary power of life is nothing more than the consequence of its fundamental properties, just as the development of structures and behaviors and the apparent purposefulness of organic life represent the unfolding of what might be thought of as the longing of the first living organism—to reproduce. Life is a process, one that is capable of generating new properties. For just this reason it is impossible to give a fixed, unchanging definition. The solution to the problem that I have proposed does not amount to a list of all the relevant characteristics of life, past, present, and future; it is a group of three fundamental properties, or pillars as I have called them (rather than principles), which, taken together, are responsible for the appearance of all other characteristics.

Not so long ago, philosophers and scientists spoke of the "dialectical power" of life as the source of new properties. Today, we say that life generates "emergent" phenomena. The difference in these two terms cannot conceal the fact that they are meant to describe the same observations; indeed, they are extremely similar, for both seek to account for the appearance of novelty in the world—of phenomena that were not present at the moment of its original creation, as it were, but that represent a kind of continual, open-ended inventiveness. It is possible that the emergence of such complex properties—for example, the appearance of cognitive capacities—is a rare phenomenon, due to the conjunction of exceedingly unlikely events; or it may be relatively frequent and so, in principle at least, observable in all possible living worlds. As we have seen, there is little agreement on this point.

It might be argued, of course, that by regarding life as a process of perpetual transformation it becomes impossible to show that it ever appeared at all. The notion of process in this sense raises the problem of how to differentiate between life and non-life, and suggests that evolution has deep, unmapped roots in the inanimate world. It may be helpful to consider a comparable difficulty. A few decades ago, no one would have hesitated for a moment in drawing up a list of the characteristics that distinguish humans from animals: language, tool use, consciousness of death, religious feelings, transmission of a shared culture. Nearly all these clear lines of demarcation have since become blurred. We now know that apes can acquire a form of language, manipulate tools, and transmit their cultural habits. It seems probable, too, that many animals also have a certain awareness of death.[4] These differences, once confidently assumed to be qualitative, are now seen to be quantitative, a matter of degree. Must we therefore abandon all attempts to differentiate humans and animals—or life from non-life? Perhaps some objects are simply more alive than others, just as some animals are more "human" than others. Trying to identify the simplest organism may therefore turn out to be as pointless as trying to identify the smallest giant.[5] Paradoxically, just when the dangers of human activity for the natural world are becoming plain, the differences between human beings and other forms of life are no longer as clear as they once seemed to be.

The problem in each case arises from a failure to see that a qualitative difference can emerge from a process of perpetual evolution—that life can come from non-life, just as human thought developed from animal behavior.[6] A small step in the right direction would be to reject two familiar and superficially attractive solutions, each of which threatens ultimately to paralyze the future advance of scientific knowledge. Both of these solutions, although they are opposed to each other, deny the existence of novelty. The first is the philosophical doctrine of preformation, which holds that everything in the world has already been present from the beginning, and dismisses the role of history and the possibility of de novo creation. The logical—or perhaps pathological—extension of this idea is the so-called anthropic principle, according to which the characteristics of the universe and the values of its fundamental physical constants are explained by the need to create human beings. The second solution, endorsed by many scientists, is to say that no presently observable phenomena—including life itself—are really new, because they do not differ in any fundamental way from what came before: the history of the world is characterized by continuity rather than discontinuity. But this is mistaken. Continuity in historical processes does not preclude the existence of critical thresholds, which, once exceeded, yield qualitatively novel phenomena. Nor does it offend against rationality to assert as much!

The problem of life has always been a challenge for scientists, whose attitudes oscillate between two extremes: on the one hand, denying the existence of original characteristics peculiar to organisms, like the iatromechanists who came after Descartes and Galileo in the seventeenth century and who claimed to explain natural phenomena on the basis of evidently insufficient scientific knowledge; and, on the other, accepting the existence of such characteristics, like the vitalists, while insisting that they lie beyond the reach of naturalistic explanation, which, though it may account for the present, cannot predict the future. The proper course lies between these two extreme positions, recognizing the distinctive character of the phenomena of the living world without renouncing the attempt to explain them by natural means.

It may be instructive to examine once again the conclusions arrived at by Claude Bernard, who addressed the question "What is life?" more directly than most biologists of his time. In *Phenomena of Life Common to Animals and to Plants* (1878), Bernard drew up a list of five general characteristics of organic life: organization, generation, nutrition, evolution, and decay, sickness, and death. Of these, evolution (what today we would call embryological development) seemed to him the most remarkable, and therefore the distinguishing feature of life itself. At first glance, this answer comes as a disappointment. The characteristics I have put forward correspond more or less precisely to the first three on his list. Are we to suppose, then, that more than a century has passed without any real progress having been made? One has the impression that, among a small group of characteristics that have long been agreed to be typical of life, there is only an incessant process of shuffling and rearrangement, so that what once was to be explained becomes the fundamental principle that explains the others, and vice versa. What Bernard regarded as the key to life—development—is now seen as an indirect consequence of three other characteristics, which in turn are considered fundamental.

But on closer examination one discovers that there are a number of important differences between the two sets of answers, differences that are significant enough not only to remove any lingering doubts about the progress that has been made in the past hundred and twenty-five years, but also to show that we do now in fact know what life is. Modern biologists have succeeded in characterizing the totality of exchanges between living organisms and their environment. What Bernard called organization is based mainly on the structural properties of macromolecules and their interrelations, which have now been precisely described. Nutrition (or, as we would say, metabolism) is the sum of chemical reactions that permit the renewal of these macromolecular structures. Furthermore, we are now able to show that generation (what we would call reproduction), organization, and nutrition are closely related phenomena, and the nature of the links between them is at last fairly well understood. On the other hand, as we have seen, how these links were first established, and what form they originally assumed, remain largely unknown.

We are now therefore in a position to answer the question "What is life?" by appealing, not to some general principle, but to three fundamental characteristics. What is more, we are able to distinguish actual organisms from artificial creatures that imitate certain vital characteristics without, however, possessing the material properties that allow life to exist. Even so, we should not attempt to hide our ignorance. The systemic view of the living world is still in its infancy, and further research on the relations between macromolecules will undoubtedly produce many surprises. But no matter that the road to life may still be only very incompletely charted, the nature of its final destination is no longer an enigma.

Notes

Part I: The Death and Resurrection of Life

1. Ernest Kahane, *La vie n'existe pas* (Paris: Éditions Rationalistes, 1962); Stanley Shostak, *Death of Life: The Legacy of Molecular Biology* (London: Macmillan, 1998); François Jacob, *The Logic of Life: A History of Heredity,* trans. Betty E. Spillmann (Princeton: Princeton University Press, 1993).

2. Vitalism in its modern form is the doctrine, formulated in the eighteenth century by the German chemist Georg Ernst Stahl (1660–1734), according to which the properties of organisms are not reducible to the laws of physics and chemistry and can be explained only by the existence of a vital force.

CHAPTER ONE: THE TWILIGHT OF LIFE

1. Cited in L. E. Kay, "W. M. Stanley's crystallization of the tobacco mosaic virus, 1930–1940," *Isis* 77 (1986): 450–472; see also Angela N. H. Creager, *The Life of a Virus: Tobacco Mosaic Virus as an Experimental Model* (Chicago: University of Chicago Press, 2002).

2. Michel Morange, *A History of Molecular Biology,* trans. Matthew Cobb (Cambridge, Mass.: Harvard University Press, 1998), 235–237.

3. Georges Canguilhem, "The Vitalist Imperative," in *A Vital Rationalist: Selected Writings from Georges Canguilhem,* ed. François Delaporte, trans. Arthur Goldhammer (New York: Zone, 1994), 287–291.

4. Wolfgang Ostwald, *Introduction to Theoretical and Applied Colloid Chemistry: The World of Neglected Dimensions,* trans. Martin H. Fischer (New York: John Wiley and Sons, 1917).

5. S. Podolsky, "The role of the virus in origin-of-life theorizing," *J. Hist. Biol.* 29 (1996): 79–126; William C. Summers, *Félix d'Herelle and the Origins of Molecular Biology* (New Haven: Yale University Press, 1999); Creager, *The Life of a Virus.*

6. Michel Morange, *The Misunderstood Gene,* trans. Matthew Cobb (Cambridge, Mass.: Harvard University Press, 2001), 11–17.

7. Leonard Troland was the first to identify life with the power of genes, which he conceived as analogous to the catalytic power of enzymes; L. T. Troland, "Biological enigmas and the theory of enzyme action," *The American Naturalist* 51 (1917): 321–350; also A. W. Ravin, "The gene as catalyst; the gene as organism," *Stud. Hist. Biol.* 1 (1977): 1–45.

8. J. Alexander and C. B. Bridge, "Some physiochemical aspects of life, mutation, and evolution," *Science* 70 (1929): 508–510.

9. H. J. Muller, "The Gene as the Basis of Life," in *Proceedings of the International Congress of Plant Science, Ithaca, New York, August 16–23, 1926,* ed. B. M. Duggar, 2 vols. (Menosha, Wis.: G. Banta, 1929), 1:897–921.

10. H. J. Muller, "Variation due to change in the individual gene," *The American Naturalist* 56 (1922): 32–50.

11. M. Delbrück, "Preliminary write-up on the topic 'Riddle of Life,'" in "A physicist's renewed look at biology: Twenty years later," *Science* 168 (1970): 1314–1315.

12. R. G. Green, "On the nature of filterable viruses," *Science* 82 (1935): 443–445.

13. This distinction is discussed in Freeman Dyson, *Origins of Life,* 2nd rev. ed. (Cambridge: Cambridge University Press, 1999), 6–8; see also Richard Dawkins, *The Selfish Gene* (Oxford: Oxford University Press, 1976).

14. Robert Bud, *The Uses of Life: A History of Biotechnology* (Cambridge: Cambridge University Press, 1993).

CHAPTER TWO: LIFE AS GENETIC INFORMATION

1. J. Cello, A. V. Paul, and E. Wimmer, "Chemical synthesis of poliovirus cDNA: Generation of infectious virus in the absence of natural template," *Science* 297 (2002): 1016–1018.

2. Erwin Schrödinger, *What Is Life? The Physical Aspect of the Living Cell* (Cambridge: Cambridge University Press, 1944; reprinted with foreword by Roger Penrose, 1992), 3–5.

3. A. Lwoff, "The concept of virus," *J. Gen. Microbiol.* 17 (1957): 239–253.

4. Georges Canguilhem, "The Concept of Life," in *A Vital Rationalist:*

Selected Writings from Georges Canguilhem, ed. François Delaporte (New York: Zone, 1994), 303–319.

5. A. I. Oparin, *The Origin of Life* (1924; first English translation 1938), and J. B. S. Haldane, "The Origin of Life" (1929), reprinted in *Origin of Life,* ed. J. D. Bernal (London: Weidenfeld and Nicolson, 1967), 199–234, 242–249.

6. S. L. Miller, "Production of amino acids under possible primitive Earth conditions," *Science* 117 (1953): 528.

7. John von Neumann, "The General and Logical Theory of Automata," reprinted in *Cerebral Mechanisms in Behavior,* ed. Lloyd A. Jeffress (New York: John Wiley and Sons, 1951), 1–41.

8. For a critical history of cybernetics and its complicated legacy for contemporary science and philosophy, see Jean-Pierre Dupuy, *On the Origins of Cognitive Science: The Mechanization of the Mind,* trans. M. B. DeBevoise (Cambridge, Mass.: The MIT Press, 2008).

CHAPTER THREE: THE RETURN OF LIFE

1. Carl R. Woese, *The Genetic Code: The Molecular Basis for Genetic Expression* (New York: Harper and Row, 1967); F. H. C. Crick, "The origin of the genetic code," *J. Mol. Biol.* 38 (1968): 367–379; L. E. Orgel, "Evolution of the genetic apparatus," *J. Mol. Biol.* 38 (1968): 381–393.

2. N. R. Pace and T. L. Marsh, "RNA catalysis and the origin of life," *Orig. Life* 16 (1985): 97–116.

3. W. Gilbert, "The RNA world," *Nature* 319 (1986): 319, 618.

4. L. E. Orgel, "Prebiotic chemistry and the origin of the RNA world," *Crit. Rev. Biochem. Mol. Biol.* 39 (2004): 99–123.

5. A. G. Cairns-Smith, "The origin of life and the nature of the primitive gene," *J. Theoret. Biol.* 10 (1965): 53–88.

6. The ability of proteins to change the form of other proteins (the source, for example, of "prion" diseases) does not alter the fact that only nucleic acids can be informational in the second sense of the word.

7. There are still some scientists who think that the principle of life can be found in molecular information. This approach tends to be similar to certain theories of complexity (discussed in Chapter 15). See, for example, Werner R. Loewenstein, *The Touchstone of Life: Molecular Information, Cell Communication, and the Foundations of Life* (New York: Oxford University Press, 1999).

8. Evelyn Fox Keller, *The Century of the Gene* (Cambridge, Mass.: Har-

vard University Press, 2000), 80–82. At the October 2003 Ig Nobel Prize ceremony (a send-up of the actual Nobel Prizes), the geneticist Eric Lander facetiously summarized the Human Genome Project in seven words: "Genome. Bought the book. Hard to read." See S. Nadis, "Spoof Nobels take researchers for a ride," *Nature* 425 (2003): 550.

9. Morange, *The Misunderstood Gene*, 64–82.

10. F. Jacob and J. Monod, "Genetic regulatory mechanisms in the synthesis of proteins," *J. Mol. Biol.* 3 (1961): 318–356. The notion of a genetic program was simultaneously introduced by Ernst Mayr.

11. Henri Atlan, *La fin du "tout génétique"? Vers de nouveaux paradigmes en biologie* (Paris: INRA Éditions, 1999), 23–25; see also Keller, *The Century of the Gene*, 80–87.

12. M. Morange, "Gene function," *C. R. Acad. Sci. Paris, Sciences de la vie* 323 (2000): 1147–1153.

13. B. Cunningham, "The reemergence of 'emergence,'" *Philosophy of Science* 68 (2001): S62–S75.

14. Daniel Andler, Anne Fagot-Largeault, and Bertrand Saint-Sernin, *Philosophie des Sciences* (Paris: Gallimard, 2002), chapter 8.

15. S. J. Dick, "NASA and the search for life in the universe," *Endeavour* 30 (2006): 71–75; and Steven J. Dick and James L. Strick, *The Living Universe: NASA and the Development of Astrobiology* (New Brunswick, N.J.: Rutgers University Press, 2004).

16. J. Lederberg, "Exobiology: Approaches to life beyond the Earth," *Science* 132 (1960): 393–400; see also M. Morange, "Fifty years ago: The beginnings of exobiology," *J. Biosci.* 32 (2007): 1083–1087.

17. Dick and Strick, *The Living Universe*, 80–102.

18. H. Gavaghan, "ESA embraces astrobiology," *Science* 292 (2001): 1626–1627.

19. J. J. Lissauer, "Extrasolar planets," *Nature* 419 (2002): 355–358; C. Lovis, M. Mayor, F. Pepe, et al., "An extrasolar planetary system with three Neptune-mass planets," *Nature* 441 (2006): 305–309.

20. W. Cash, "Detection of Earth-like planets around nearby stars using a petal-shaped occulter," *Nature* 442 (2006): 51–53.

21. G. Schilling, "Habitable, but not much like home," *Science* 316 (2007): 528.

Part II: The Question in Historical Perspective

1. Gerald Holton, *The Scientific Imagination: Case Studies* (Cambridge: Cambridge University Press, 1978), 3–24.

2. Georges Canguilhem, "Aspects du vitalisme," in *La connaissance de la vie* (Paris: Vrin, 1975), 85.

3. Maurice Merleau-Ponty, *Le visible et l'invisible* (Paris: Gallimard, 1964), 47.

CHAPTER FOUR: A RICH HERITAGE

1. Aristotle, *De Anima* 2.1.14–15, in *The Basic Works of Aristotle*, ed. Richard McKeon (New York: Random House, 1941), 555.

2. Xavier Bichat, *Recherches physiologiques sur la vie et la mort*, ed. André Pichot (Paris: Flammarion, 1994).

3. Michel Foucault, *The Order of Things: An Archaeology of the Human Sciences* (New York: Pantheon, 1971), 126–128.

4. Troland, "Biological enigmas and the theory of enzyme action."

5. Schrödinger, *What Is Life?* 56–66.

6. F. Jacob, "The Leeuwenhoek Lecture, 1977: Mouse teratocarcinoma and mouse embryo," *Proc. Roy. Soc. London Ser. B.*, 201 (1978): 249–270.

7. For a detailed survey of these various scientific and philosophical conceptions, see André Pichot, *Histoire de la notion de vie* (Paris: Gallimard, 1993).

8. Félix d'Herelle, *Les défenses de l'organisme* (Paris: Flammarion, 1923).

CHAPTER FIVE: CONTEMPORARY ANSWERS

1. D. E. Koshland, "The seven pillars of life," *Science* 295 (2002): 2215–2216.

2. Friedrich Engels, *Anti-Dühring: Herr Eugen Dühring's Revolution in Science* (Moscow: Progress, 1947), 46. My attention was drawn to this quote by Pier Luigi Luisi, "About various definitions of life," *Orig. Life Evol. Biosph.* 28 (1998): 613–622.

3. From Joyce's foreword to David W. Deamer and Gail R. Fleischaker, eds., *Origins of Life: The Central Concepts* (Boston: Jones and Bartlett, 1994), xi–xii; see also Luisi, "About various definitions of life."

4. Patrice David and Sarah Samadi, *La théorie de l'évolution: Une logique pour la biologie* (Paris: Flammarion, 2000), 7–10.

5. This recalls a distinction made earlier by the Hungarian chemical engineer Tibor Gánti between "potential" and "absolute" defining criteria of

life; see *The Principles of Life* (Oxford: Oxford University Press, 2003; originally published in Budapest in 1971).

6. Claude Bernard, *Phenomena of Life Common to Animals and to Plants,* trans. R. P. Cook and M. A. Cook (Dundee: Cook & Cook, 1974).

7. Julio Fernández Ostolaza and Álvaro Moreno Bergareche, *Vida artificial* (Madrid: Eudema, 1992).

8. Christopher G. Langton, ed., *Artificial Life,* vol. 6 of *Santa Fe Studies in the Sciences of Complexity* (Reading, Mass.: Addison-Wesley, 1989).

9. J. Perrett, "Biochemistry and bacteria," *New Biology* 12 (1952): 68–69.

10. F. J. Varela, H. R. Maturana, and R. Uribe, "Autopoiesis: The organization of living systems, its characterization and a model," *BioSystems* 5 (1974): 187–196; Francisco J. Varela, *Principles of Biological Autonomy* (New York: Elsevier, 1979), G. R. Fleischaker, "Origins of life: An operational definition," *Orig. Life Evol. Biosph.* 20 (1990): 127–137.

11. Harold J. Morowitz, *Beginnings of Cellular Life: Metabolism Recapitulates Biogenesis* (New Haven: Yale University Press, 1992), 4–6. The other definitions I have mentioned were proposed after the publication of this book, but Morowitz is not known to have changed his position in the interval.

12. Some of these definitions have since been modified—and improved—by their authors. See, for instance, K. Ruiz-Mirazo, J. Pereto, and Á. Moreno, "A universal definition of life: Autonomy and open-ended evolution," *Orig. Life Evol. Biosph.* 34 (2004): 323–346.

Part III: Contributions of Current Research

CHAPTER SIX: LOOKING FOR LIFE'S PAST

1. W. Thomson, Presidential Address to the British Association for the Advancement of Science, *Nature* 4 (1871): 262; see also S. Arrhenius, "The propagation of life in space," *Die Umschau* 7 (1903): 481.

2. See the untitled article by P. Becquerel, *Bull. Soc. Astron.* 38 (1924): 393.

3. J. Oro, "Comets and the formation of biochemical compounds on the primitive Earth," *Nature* 190 (1961): 389–390.

4. P. Ehrenfreund, D. P. Glavin, O. Botta, et al., "Extraterrestrial amino acids in Orgueil and Ivuna: Tracing the parent body of CI type carbonaceous chondrites," *Proc. Natl. Acad. Sci. USA* 98 (2001): 2138–2141.

5. F. Crick and L. E. Orgel, "Directed panspermia," *Icarus* 19 (1973): 341–346.

6. See, for example, Fred Hoyle and Chandra Wickramasinghe, *Our Place in the Cosmos: The Unfinished Revolution?* (London: Phoenix, 1996).

7. D. S. McKay, E. K. Gibson, Jr., K. L. Thomas-Keprta, et al., "Search for past life on Mars: Possible relic biogenic activity in Martian meteorite ALH84001," *Science* 273 (1996): 924–930.

8. See the many objections to these arguments published in the 20 September 1996 and 20 December 1996 issues of *Science;* for a journalistic account of the whole affair, see Kathy Sawyer, *The Rock from Mars: A Detective Story on Two Planets* (New York: Random House, 2006).

9. On the origin and development of terrestrial life generally, see Freeman Dyson, *Origins of Life* (Cambridge: Cambridge University Press, 2nd rev. ed., 1999); Iris Fry, *The Emergence of Life on Earth: A Historical and Scientific Overview* (New Brunswick, N.J.: Rutgers University Press, 2000); Christian de Duve, *Life Evolving: Molecules, Mind, and Meaning* (Oxford: Oxford University Press, 2002); J. William Schopf, ed., *Life's Origin: The Beginning of Biological Evolution* (Berkeley: University of California Press, 2002); Christian de Duve, *Singularities: Landmarks on the Pathways of Life* (Cambridge: Cambridge University Press, 2005); Robert M. Hazen, *Genesis* (Washington, D.C.: Joseph Henry Press, 2005); and Pier Luigi Luisi, *The Emergence of Life: From Chemical Origins to Synthetic Biology* (Cambridge: Cambridge University Press, 2006).

10. J. William Schopf, *Cradle of Life: The Discovery of Earth's Earliest Fossils* (Princeton: Princeton University Press, 1999); J. W. Schopf, "Microfossils of the Early Archaean Apex chert: New evidence of the antiquity of life," *Science* 260 (1993): 640–646; J. W. Schopf, A. B. Kudryavtsev, D. G. Agresti, et al., "Laser-Raman imagery of Earth's earliest fossils," *Nature* 416 (2002): 73–76.

11. See M. D. Brasier, O. R. Green, A. P. Jephcoat, et al., "Questioning the evidence for Earth's oldest fossils," *Nature* 416 (2002): 76–81; C. M. Fedo and M. J. Whitehouse, "Metasomatic origin of quartz-pyroxene rock, Akilia, Greenland, and implications for Earth's earliest life," *Science* 296 (2002): 1448–1452; M. A. van Zuilen, A. Lepland, and G. Arrhenius, "Reassessing the evidence for the earliest traces of life," *Nature* 418 (2002): 627–630; and J. M. García-Ruiz, S. T. Hyde, A. M. Carnerup, et al., "Self-assembled silica-carbonate structures and detection of ancient microfossils," *Science* 302 (2003): 1194–1197.

12. M. M. Tice and D. R. Lowe, "Photosynthetic microbial mats in the 3,416-Myr-old ocean," *Nature* 431 (2004): 549–552; H. Furnes, N. R. Banerjee, K. Muehlenbachs, et al., "Early life recorded in Archaean pillow lavas," *Science* 304 (2004): 578–581; Y. Ueno, K. Yamada, N. Yoshida, et al., "Evidence from fluid inclusions for microbial methanogenesis in the early Archaean

era," *Nature* 440 (2006): 516–519; and A. C. Allwood, M. R. Walter, B. S. Kamber, C. P. Marshall, and I. W. Burch, "Stromatolites reef from the early Archaean era of Australia," *Nature* 441 (2006): 714–718.

13. G. F. Joyce, "The antiquity of RNA-based evolution," *Nature* 418 (2002): 214–221.

14. For an opposite point of view see Simon Conway Morris, *Life's Solution: Inevitable Humans in a Lonely Universe* (New York: Cambridge University Press, 2003).

15. Peter D. Ward and Donald Brownlee, *Rare Earth: Why Complex Life Is So Uncommon in the Universe* (New York: Copernicus, 2000); B. W. Aldiss, "Desperately seeking aliens," *Nature* 409 (2001): 1080–1082; S. B. Carroll, "Chance and necessity: The evolution of morphological complexity and diversity," *Nature* 409 (2001): 1102–1109; Mary Shelley, *Frankenstein, or, The Modern Prometheus,* ed. M. K. Joseph (New York: Oxford University Press, 1998). Given that the existence of spontaneous generation on the basis of compounds produced by organic decomposition was widely accepted in Shelley's time, it would be more exact to say that the barrier lay between components from the living world and components that are no longer part of that world.

16. C. R. Woese, "Default taxonomy: Ernst Mayr's view of the microbial world," *Proc. Natl. Acad. Sci. USA* 95 (1998): 11043–11046.

17. S. d'Hondt, B. B. Jorgensen, D. J. Miller, et al., "Distributions of microbial activities in deep subseafloor sediments," *Science* 306 (2004): 2216–2221.

18. Stephen Jay Gould, *Wonderful Life: The Burgess Shale and the Nature of History* (New York: W. W. Norton, 1989), 55–60; see also A. H. Knoll and S. B. Carroll, "Early animal evolution: Emerging views from comparative biology and geology," *Science* 284 (1999): 2129–2137.

19. Only 5 percent of species are thought to have survived the extinction event at the Permo-Triassic boundary, 250 million years ago. See Douglas H. Erwin, *Extinction: How Life on Earth Nearly Ended 250 Million Years Ago* (Princeton: Princeton University Press, 2006).

20. P. G. Falkowski, M. E. Katz, A. J. Milligan, et al., "The rise of oxygen over the past 205 million years and the evolution of large placental mammals," *Science* 309 (2005): 2202–2204; see also M. Cardillo, G. M. Mace, K. E. Jones, et al., "Multiple causes of high extinction risk in large mammal species," *Science* 309 (2005): 1239–1241.

21. For opposing views, see Stephen Webb, *If the Universe Is Teeming with Aliens . . . Where Is Everybody? Fifty Solutions to the Fermi Paradox and the Problem of Extraterrestrial Life* (New York: Copernicus, 2002).

CHAPTER SEVEN: RETRACING THE PATH OF LIFE

1. See, for instance, M. Klussmann, H. Iwamura, S. P. Mathew, et al., "Thermodynamic control of asymmetric amplification in amino acid catalysis," *Nature* 441 (2006): 621–623.

2. R. Shapiro, "Prebiotic ribose synthesis: A critical analysis," *Orig. Life Evol. Biosph.* 18 (1988): 71–85.

3. L. Orgel, "A simpler nucleic acid," *Science* 290 (2000): 1306–1307.

4. For a vivid description of this debate see Robert M. Hazen, *Genesis* (Washington, D.C.: Joseph Henry Press, 2005).

5. G. Wächterhäuser, "Evolution of the first metabolic cycles," *Proc. Natl. Acad. Sci. USA* 87 (1990): 200–204.

6. G. D. Cody, "Transition metal sulfides and the origin of metabolism," *Ann. Rev. Earth Planetary Sci.* 32 (2004): 569–599.

7. J. W. Szostak, D. P. Bartel, and P. L. Luisi, "Synthesizing life," *Nature* 409 (2001): 387–390; W. K. Johnston, P. J. Unrau, M. S. Lawrence, et al., "RNA-catalyzed RNA polymerization: Accurate and general RNA-templated primer extension," *Science* 292 (2001): 1319–1325; K. E. McGinness and G. F. Joyce, "In search of an RNA replicase ribozyme," *Chem. Biol.* 10 (2003): 5–14; and M. S. Lawrence and D. P. Bartel, "New ligase-derived RNA polymerase ribozymes," *RNA* 11 (2005): 1173–1180.

8. P. Szabo, I. Scheuring, T. Czaran, and E. Szathmary, "*In silico* simulations reveal that replicators with limited dispersal evolve towards higher efficiency and fidelity," *Nature* 420 (2002): 340–343.

9. Manfred Eigen was among the first to have shown by simulation that such optimization processes are possible in M. Eigen, W. Gardiner, P. Schuster, and R. Winkler-Oswatitsch, "The origin of genetic information," *Sci. Am.* 244 (1981): 78–94; see also M. Eigen, C. K. Biebricher, M. Gebinoga, and W. C. Gardiner, "The hypercycle: Coupling of RNA and protein biosynthesis in the infection cycle of an RNA bacteriophage," *Biochemistry* 30 (1991): 11005–11018.

10. S. Fusz, A. Elsenführ, S. G. Srivatsan, A. Heckel, and M. Famulok, "A ribozyme for the aldol reaction," *Chemistry and Biology* 12 (2005): 941–950.

11. Eigen et al., "The hypercycle."

12. This distinction was proposed two decades ago by the Cambridge historian of science Harmke Kamminga in "Historical perspective: The problem of the origin of life in the context of developments in biology," *Orig. Life Evol. Biosph.* 18 (1988): 1–11; see, too, the recent discussion in Christopher Wills and Jeffrey Bada, *The Spark of Life: Darwin and the Primeval Soup* (New York: Perseus, 2000), xv–xix.

13. See Part 2 of Stuart A. Kauffman, *The Origin of Order: Self-Organization and Selection in Evolution* (New York: Oxford University Press, 1993), 287–404.

14. D. Segré, D. Ben-Eli, D. W. Deamer, and D. Lancet, "The lipid world," *Orig. Life Evol. Biosph.* 31 (2001): 119–145.

15. See Chapter 6, note 9.

16. I. A. Chen, R. W. Roberts, and Jack W. Szostak, "The emergence of competition between model protocells," *Science* 305 (2004): 1474–1476.

17. The American biochemist Sidney W. Fox's synthesis of proteinoids—amino acid polymers whose wonderful catalytic and structural properties, more imaginary than real as it turned out, were supposed to readily explain the transition from non-life to life—remains a cautionary lesson for everyone working on this problem; see S. W. Fox, "Spontaneous generation, the origin of life, and self-assembly," *Curr. Mod. Biol.* 2 (1968): 235–240.

18. Robert Shapiro, *Origins: A Skeptic's Guide to the Creation of Life on Earth* (New York: Summit, 1986).

19. M. M. Hanczyc, S. M. Fujikawa, and J. W. Szostak, "Experimental models of primitive cellular compartments: encapsulation, growth, and division," *Science* 302 (2003): 618–622.

CHAPTER EIGHT: READING THE PALIMPSEST OF LIFE

1. S. M. Siegel, K. Roberts, H. Nathan, and O. Daly, "Living relative of the microfossil *Kakabekia*," *Science* 158 (1967): 1231–1234.

2. S. A. Benner, A. D. Ellington, and A. Tauer, "Modern metabolism as a palimpsest of the RNA world," *Proc. Natl. Acad. Sci. USA* 86 (1989): 7054–7058.

3. T. R. Cech, "The ribosome is a ribozyme," *Science* 289 (2000): 878–879.

4. W. Winkler, A. Nahvi, and R. R. Breaker, "Thiamine derivatives bind messenger RNAs directly to regulate bacterial gene expression," *Nature* 419 (2002): 952–956.

5. C. Huber and G. Wächtershäuser, "α-Hydroxy and α-amino acids under possible Hadean, volcanic origin of life conditions," *Science* 314 (2006): 630–632.

6. C. Cunchillos and G. Lecointre, "Evolution of amino acid metabolism inferred through cladistic analysis," *J. Biol. Chem.* 278 (2003): 47960–47970; C. Cunchillos and G. Lecointre, "Integrating the universal metabolism into a phylogenetic analysis," *Mol. Biol. Evol.* 22 (2005): 1–11.

7. S. J. Freeland, R. D. Knight, and L. F. Landweber, "Do proteins predate DNA?" *Science* 286 (1999): 690–692.

8. C. R. Woese, "Order in the genetic code," *Proc. Natl. Acad. Sci. USA* 54 (1965): 71–75; C. R. Woese, "On the evolution of the genetic code," *Proc. Natl. Acad. Sci. USA* 54 (1965): 1546–1552; J. T.-F. Wong, "A co-evolution theory of the genetic code," *Proc. Natl. Acad. Sci. USA* 72 (1975): 1909–1912; Eigen et al., "The origin of genetic information"; T. A. Ronneberg, L. F. Landweber, and S. J. Freeland, "Testing a biosynthetic theory of the genetic code: Fact or artefact?" *Proc. Natl. Acad. Sci. USA* 97 (2000): 13690–13695; V. Pezo, D. Metzgar, T. L. Hendrickson, et al., "Artificially ambiguous genetic code confers growth yield advantage," *Proc. Natl. Acad. Sci. USA* 101 (2004): 8593–8597; and M. Yarus, J. G. Caporaso, and R. Knight, "Origins of the genetic code: The escaped triplet theory," *Annu. Rev. Biochem.* 74 (2005): 179–198.

9. E. A. Gaucher, J. M. Thomson, M. F. Burgan, and S. A. Benner, "Inferring the palaeoenvironment of ancient bacteria on the basis of resurrected proteins," *Nature* 425 (2003): 285–288.

CHAPTER NINE: LIFE UNDER EXTREME CONDITIONS

1. L. J. Rothschild and R. L. Mancinelli, "Life in extreme environments," *Nature* 409 (2001): 1092–1100; see also David A. Wharton, *Life at the Limits: Organisms in Extreme Environments* (Cambridge: Cambridge University Press, 2002).

2. P. Forterre, "A hot topic: The origin of hyperthermophiles," *Cell* 85 (1996): 789–792; N. Galtier, N. Tourasse, and M. Gouy, "A nonhyperthermophilic common ancestor to extant life-forms," *Science* 283 (1999): 220–221; J. L. Bada and A. Lazcano, "Some like it hot, but not the first biomolecules," *Science* 296 (2002): 1982–1983.

3. Keller, *Century of the Gene,* 13–25.

4. For a different point of view on the ancestry of reverse gyrase, see E. Waters, M. J. Hohn, I. Ahel, et al., "The genome of *Nanoarchaeum equitans:* Insights into early archaeal evolution and derived parasitism," *Proc. Natl. Acad. Sci. USA* 100 (2003): 12984–12988.

5. R. F. Service, "Creation's seventh day," *Science* 289 (2000): 232–235; L. Wang, A. Brock, B. Herberich, et al., "Expanding the genetic code of *Escherichia coli,*" *Science* 292 (2001): 498–500; V. Döring, H. D. Mootz, L. A. Nangle, et al., "Enlarging the amino acid set of *Escherichia coli* by infiltration of the valine coding pathway," *Science* 292 (2001): 501–504; I. Hirao, T. Ohtsuki, T. Fujiwara, et al., "An unnatural base pair for incorporating amino acid analogs into proteins," *Nature Biotechnology* 20 (2002): 177–182.

6. J. F. Atkins and R. Gesteland, "The 22nd amino acid," *Science* 296 (2002): 1409–1410.

CHAPTER TEN: THE SEARCH FOR A MINIMAL GENOME

1. Jean-Baptiste de Lamarck, *Zoological Philosophy: An Exposition with Regard to the Natural History of Animals,* trans. Hugh Elliot (London: Hafner, 1963), 2.

2. Rowland H. Davis, *The Microbial Models of Molecular Biology: From Genes to Genomes* (Oxford: Oxford University Press, 2003), 80–98.

3. H. Huber, M. J. Hohn, R. Rachel, et al., "A new phylum of archaea represented by a nanosized hyperthermophilic symbiont," *Nature* 417 (2002): 63–67; see also Waters, Hohn, Ahel, et al., "The genome of *Nanoarchaeum equitans.*"

4. E. O. Kajander and N. Ciftcioglu, "Nanobacteria: An alternative mechanism for pathogenic intra- and extracellular calcification and stone formation," *Proc. Natl. Acad. Sci. USA* 95 (1998): 8274–8279.

5. C. M. Fraser, J. D. Gocayne, O. White, et al., "The minimal gene complement of *Mycoplasma genitalium,*" *Science* 270 (1995): 397–403.

6. Quoted in M. K. Cho, D. Magnus, A. L. Caplan, et al., "Ethical considerations in synthesizing a minimal genome," *Science* 286 (1999): 2087–2090.

7. C. A. Hutchison III, S. N. Peterson, S. R. Gill, et al., "Global transposon mutagenesis and a minimal mycoplasma genome," *Science* 286 (1999): 2165–2169.

8. A. R. Mushegian and E. V. Koonin, "A minimal gene set for cellular life derived by comparison of complete bacterial genomes," *Proc. Natl. Acad. Sci. USA* 93 (1996): 10268–10273.

9. S. Shigenobu, H. Watanabe, M. Hatori, et al., "Genome sequence of the endocellular bacterial symbiont of aphids *Buchnera* sp. APS," *Nature* 407 (2000): 81–86; see also P. R. Gilson, V. Su, C. H. Slamovits, et al., "Complete nucleotide sequence of the chlorarachniophyte nucleomorph: Nature's smallest nucleus," *Proc. Natl. Acad. Sci. USA* 103 (2006): 9566–9571.

10. R. Gil, F. J. Silva, J. Pereto, and A. Moya, "Determination of the core of a minimal bacterial gene set," *Microbiol. Mol. Biol. Rev.* 68 (2004): 518–537.

11. André Lwoff, *L'évolution physiologique: Étude des pertes de fonction chez les microorganismes* (Paris: Hermann, 1944).

12. See note 3 above.

13. This idea was first put forward more than a half century ago:

"Life . . . is a system of interrelations; each organism is affected by, even if not wholly dependent on, others. It is this fact more than any other that robs present-day biochemistry of any strict relevance to the problem of the origin and essential nature of life." See N. W. Pirie, "The nature and development of life," *Modern Quarterly* 3 (1948): 82–93.

14. M. Huynen, "Constructing a minimal genome," *Trends in Genetics* 16 (2000): 116. The diversity of possible solutions is also suggested by a recent modeling study, C. Pal, B. Papp, M. J. Lercher, et al., "Chance and necessity in the evolution of minimal metabolic networks," *Nature* 440 (2006): 667–670.

CHAPTER ELEVEN: ASTROBIOLOGICAL INVESTIGATIONS

1. George Basalla, *Civilized Life in the Universe: Scientists on Intelligent Extraterrestrials* (Oxford: Oxford University Press, 2005).

2. David Grinspoon, *Lonely Planets: The Natural Philosophy of Alien Life* (New York: HarperCollins, 2004).

3. See, for example, J. L. Bada, "State-of-the-art instruments for detecting extraterrestrial life," *Proc. Natl. Acad. Sci. USA* 98 (2001): 797–800.

4. G. G. Simpson, "The nonprevalence of humanoids," *Science* 143 (1964): 769–775.

5. N. R. Pace, "The universal nature of biochemistry," *Proc. Natl. Acad. Sci. USA* 98 (2001): 805–808.

6. J. Cohen and I. Stewart, "Where are the dolphins?" *Nature* 409 (2001): 1119–1122; see also their book *Evolving the Alien: The Science of Extraterrestrial Life* (London: Ebury Press/John Wiley, 2002).

7. Nonetheless things are changing, at least in government oversight of research in the United States, where a committee established by the National Research Council to investigate the limits of organic life in planetary systems has now delivered a preliminary report that shows a new openness to the possibility of discovering exotic forms of life; see *The Limits of Organic Life in Planetary Systems* (Washington, D.C.: National Academies Press, 2007).

8. K. Biemann, J. Oro, P. Toulmin III, et al., "Search for organic and volatile inorganic compounds in two surface samples from the Chryse Planitia Region of Mars," *Science* 194 (1976): 72–76.

9. H. P. Klein, N. H. Horowitz, G. V. Levin, et al., "The Viking biological investigation: Preliminary results," *Science* 194 (1976): 99–105.

10. A. S. Yen, S. S. Kim, M. H. Hecht, et al., "Evidence that the reactivity of the Martian soil is due to superoxide ions," *Science* 289 (2000): 1909–1912.

11. T. B. McCord, G. B. Hansen, and C. A. Hibbitts, "Hydrated salt minerals on Ganymede's surface: Evidence of an ocean below," *Science* 292 (2001): 1523–1525; M. G. Kivelson, K. K. Khurana, C. T. Russell, et al., "Galileo magnetometer measurements: A stronger case for a subsurface ocean at Europa," *Science* 289 (2000): 1340–1343; J. Spencer and D. Grinspoon, "Inside Enceladus," *Nature* 445 (2007): 376–377; J.-P. Bibring, Y. Langevin, J. F. Mustard, et al., "Global mineralogical and aqueous Mars history derived from OMEGA/Mars Express data," *Science* 312 (2006): 400–404; R. P. Irwin III, T. A. Maxwell, A. D. Howard, et al., "A large paleolake basin at the head of Ma'adim Vallis, Mars," *Science* 296 (2002): 2209–2212; M. C. Malin, K. S. Edgett, L. V. Posiolova, S. M. McColley, and E. Z. N. Dobrea, "Present-day impact cratering rate and contemporary gully activity on Mars," *Science* 314 (2006): 1573–1577; M. C. Malin and K. S. Edgett, "Evidence for recent groundwater seepage and surface runoff on Mars," *Science* 288 (2000): 2330–2335; P. R. Christensen, "Formation of recent Martian gullies through melting of extensive water-rich snow deposits," *Nature* 422 (2003): 45–48; M. C. Malin and K. S. Edgett, "Evidence for persistent flow and aqueous sedimentation on early Mars," *Science* 302 (2003): 1931–1934; L. P. Knauth, D. M. Burt, and K. H. Wohletz, "Impact origin of sediments at the Opportunity landing site on Mars," *Nature* 438 (2005): 1123–1128; T. M. McCollom and B. M. Hynek, "A volcanic environment for bedrock diagenesis at Meridiani Planum on Mars," *Nature* 438 (2005): 1129–1131.

12. The possibility of life without water was nonetheless discussed in December 2003 at a meeting of the Royal Society in London.

13. D. N. Thomas and G. S. Dieckmann, "Antarctic sea ice—a habitat for extremophiles," *Science* 295 (2005): 641–644; N. H. Horowitz, R. E. Cameron, and J. S. Hubbard, "Microbiology of the dry valleys of Antarctica," *Science* 176 (1972): 242–245.

14. See C. F. Chyba, "Energy for microbial life on Europa," *Nature* 403 (2000): 381–382; C. F. Chyba and C. B. Phillips, "Possible ecosystems and the search for life on Europa," *Proc. Natl. Acad. Sci. USA* 98 (2001): 801–804.

15. C. H. Lineweaver, Y. Fenner, and B. K. Gibson, "The galactic habitable zone and the age distribution of complex life in the Milky Way," *Science* 303 (2004): 59–62.

16. G. Israël, C. Szopa, F. Raulin, et al., "Complex organic matter in Titan's atmospheric aerosols from *in situ* pyrolysis and analysis," *Nature* 438 (2005): 796–799.

17. J. E. Lovelock, "A physical basis for life detection experiments," *Nature* 207 (1965): 568–570.

18. W. E. Dietrich and J. T. Perron, "The search for a topographic signature of life," *Nature* 439 (2006): 411–418.

19. C. Sagan, W. R. Thompson, R. Carlson, et al., "A search for life on Earth from the Galileo spacecraft," *Nature* 365 (1993): 715–721.

20. This condition was insisted upon by the consortium responsible for the unfortunate Beagle 2 project: see R. A. Kerr, "Putting Martian science to the test," *Science* 301 (2003): 1832–1834.

21. After the NASA Viking mission to Mars, Carl Sagan, the astronomer and pioneering astrobiologist who was in charge of the experiments, said he had a recurring nightmare during the mission as a result of NASA's refusal to equip the lander with lights (because of weight considerations). Sagan's nightmare was that every morning Viking would beam back pictures of the Martian soil, marked with the footprints of nocturnal Martian animals that could never be seen.

22. Dick and Strick, *The Living Universe,* 131–154.

23. T. Reichhardt, "Two telescopes join hunt for ET," *Nature* 440 (2006): 853.

CHAPTER TWELVE: LIFE AS A LIVING SYSTEM

1. Jacob, *The Logic of Life,* 299–324.

2. Jan Sapp, *Evolution by Association: A History of Symbiosis* (Oxford: Oxford University Press, 1994), 165–190.

3. C. G. Kurland, L. J. Collins, and D. Penny, "Genomics and the irreducible nature of eukaryotic cells," *Science* 312 (2006): 1011–1014; see also T. M. Embley and W. Martin, "Eukaryotic evolution, changes and challenges," *Nature* 440 (2006): 623–630.

4. V. Pérez-Brocal, R. Gil, S. Ramos, A. Lamelas, M. Postigo, et al., "A small microbial genome: The end of a long symbiotic relationship?" *Science* 314 (2006): 312–313.

5. Frederic Bushman, *Lateral DNA Transfer: Mechanisms and Consequences* (Cold Spring Harbor, N.Y.: Cold Spring Harbor Laboratory Press, 2002).

6. J. B. S. Haldane, "The origins of life," *New Biology* 16 (1954): 12–27.

7. E. Denamur, G. Lecointre, P. Darlu, et al., "Evolutionary implications of the frequent horizontal transfer of mismatch repair genes," *Cell* 103 (2000): 711–721.

8. K. Vestigian, C. Woese, and N. Goldenfeld, "Collective evolution and the genetic code," *Proc. Natl. Acad. Sci. USA* 103 (2006): 10696–10701.

9. M. C. Rivera and J. A. Lake, "The ring of life provides evidence for a genome fusion origin of eukaryotes," *Nature* 431 (2004): 152–155.

10. P. Forterre, "The origin of viruses and their possible roles in major evolutionary transitions," *Virus Research* 117 (2006): 5–16; see also C. Zimmer, "Did DNA come from viruses?" *Science* 312 (2006): 870–872.

11. C. R. Woese, "On the evolution of cells," *Proc. Natl. Acad. Sci. USA* 99 (2002): 8742–8747.

12. D. Raoult, S. Audic, C. Robert, et al., "The 1.2-megabase genome sequence of mimivirus," *Science* 306 (2004): 1344–1350.

13. J.-C. Ameisen, "On the origin, evolution, and nature of programmed cell death: A timeline of four billion years," *Cell Death Differ.* 9 (2002): 367–393.

14. K.-I. Kang and M. Morange, "Succès et limites de l'étude moléculaire de la mort cellulaire programmée," *Annales d'histoire et de philosophie du vivant* 4 (2001): 159–175.

15. R. Kolter and E. P. Greenberg, "The superficial life of microbes," *Nature* 441 (2006): 300–302.

16. Ameisen, "On the origin, evolution, and nature of programmed cell death"; see also the special issue edited by I. Joint, J. A. Downie, and P. Williams, "Bacterial conversations: Talking, listening, and eavesdropping," *Phil. Trans. R. Soc. B*, 29 July 2007.

17. C. Fuqua, S. C. Winans, and E. P. Greenberg, "Census and consensus in bacterial ecosystems: The LuxR-LuxI family of quorum-sensing transcriptional regulators," *Annu. Rev. Microbiol.* 50 (1996): 727–751; see also H. Engelberg-Kulka and G. Glaser, "Addiction molecules and programmed cell death and anti-death in bacterial cultures," *Annu. Rev. Microbiol.* 53 (1999): 43–70.

Part IV: A Few Necessarily Provisional Conclusions

CHAPTER THIRTEEN: OBJECTIONS AND REPLIES

1. See Jody Hey, *Genes, Categories, and Species: The Evolutionary and Cognitive Causes of the Species Problem* (New York: Oxford University Press, 2001), and the review of this book in K. L. Shaw, "Do we need species concepts?" *Science* 295 (2002): 1238–1239; Quentin D. Wheeler and Rudolf Meier, eds., *Species Concepts and Phylogenetic Theory* (New York: Columbia University Press, 2000); also Michael Ruse, *Philosophy of Biology Today* (Albany: State University of New York Press, 1988), 51–62.

2. Alain de Libera, *La querelle des universaux: De Platon à la fin du Moyen Âge* (Paris: Seuil, 1996), 11.

3. If biology was unknown, Foucault argued, "there was a very simple reason for this: that life itself did not exist. All that existed was living beings, which were viewed through a grid of knowledge constituted by natural history." See Michel Foucault, *The Order of Things*, 127–128.

4. This objection has been further discussed in Evelyn Fox Keller, *Making Sense of Life: Explaining Biological Development with Models, Metaphors, and Machines* (Cambridge, Mass.: Harvard University Press, 2002), 290–294.

5. One eminent journal has recently taken the position that life is not a scientific concept, however; see "Meanings of 'life,'" *Nature* 447 (2007): 1031–1032.

6. François Jacob, *The Possible and the Actual* (New York: Pantheon, 1982), 9–13.

7. Charles Darwin, *The Origin of Species* (London: J. M. Dent, 1972), 455.

CHAPTER FOURTEEN: DARWINISM IN ITS PROPER PLACE

1. Niles Eldredge, *Reinventing Darwin: The Great Debate at the High Table of Evolutionary Theory* (New York: Wiley, 1995).

2. Note that Lewontin does not describe the mechanisms of heredity; all that counts is that at least part of the differential adaptation is transmitted. See R. C. Lewontin, "The units of selection," *Annu. Rev. Ecol. System* 1 (1970): 1–18.

3. Eva Jablonka and Marion J. Lamb, *Evolution in Four Dimensions: Genetic, Epigenetic, Behavioral, and Symbolic Variation in the History of Life* (Cambridge, Mass.: The MIT Press, 2005).

4. Mary Jane West-Eberhard, *Developmental Plasticity and Evolution* (Oxford: Oxford University Press, 2003); see also Marc W. Kirschner and John C. Gerhart, *The Plausibility of Life: Resolving Darwin's Dilemma* (New Haven: Yale University Press, 2005).

5. John Odling-Smee, Kevin Laland, and Marcus Feldman, *Niche Construction: The Neglected Process in Evolution* (Princeton: Princeton University Press, 2003).

6. Susan Oyama, Paul Griffiths, and Russell D. Gray, *Cycles of Contingency: Developmental Systems and Evolution* (Cambridge, Mass.: The MIT Press, 2001).

7. Jean Gayon, "From Darwin to Today in Evolutionary Biology," in

The Cambridge Companion to Darwin, ed. Jonathan Hodge and Gregory Radick (Cambridge: Cambridge University Press, 2003), 240–264.

8. Vestigian et al., "Collective evolution and the genetic code."

9. Jean-Jacques Kupiec and Pierre Sonigo, *Ni Dieu ni gène: Pour une autre théorie de l'hérédité* (Paris: Seuil, 2000).

10. G. Oster, "Darwin's motors," *Nature* 417 (2002): 25.

11. Stephen Jay Gould, "Darwinian Fundamentalism," *New York Review of Books,* 12 June 1997.

12. Jean Gayon, *Darwinism's Struggle for Survival: Heredity and the Hypothesis of Natural Selection,* trans. Matthew Cobb (Cambridge: Cambridge University Press, 1998).

13. For a totally different view, where novelties in evolution are explained by rules operating at the molecular and cellular levels, see Kirschner and Gerhart, *The Plausibility of Life.* Specialists in the currently very fashionable field of "Evo-Devo" similarly focus their research on variations in developmental genes, which they suspect have exerted a dramatic influence on evolution.

14. Orgel, "Evolution of the genetic apparatus."

15. Austin Burt and Robert Trivers, *Genes in Conflict: The Biology of Selfish Genetic Elements* (Cambridge, Mass.: Belknap Press of Harvard University Press, 2006).

16. See, for example, Lederberg, "Exobiology."

17. It is sometimes difficult to know whether one is dealing with scientific or ideological Darwinism, though the tone of ultra-Darwinist rhetoric (particularly in reverential phrases such as "the awesome power of natural selection") is usually a giveaway. See the review of Janice Moore, *Parasites and the Behavior of Animals* (Oxford: Oxford University Press, 2002), by P. Schmid-Hempel, "The awesome power of natural selection," *Nature* 417 (2002): 592.

18. Jorge Luis Borges, *The Book of Imaginary Beings,* trans. Norman Thomas di Giovanni (New York: Dutton, 1969), 15, 16.

19. D. L. Robertson and G. F. Joyce, "Selection *in vitro* of an RNA enzyme that specifically cleaves single-stranded DNA," *Nature* 344 (1990): 467–468; see also C. Tuerk and L. Gold, "Systematic evolution of ligands by exponential enrichment: RNA ligands to bacteriophage T4 DNA polymerase," *Science* 249 (1990): 505–510, and A. D. Ellington and J. W. Szostak, "*In vitro* selection of RNA molecules that bind specific ligands," *Nature* 346 (1990): 818–822.

20. Jacques Monod, *Chance and Necessity: An Essay on the Natural Philosophy of Modern Biology,* trans. Austryn Wainhouse (New York: Knopf,

1971), 165–168; see also Richard Dawkins, "Memes: The New Replicators," in *The Selfish Gene,* 189–201.

21. See, for example, R. E. Lenski, "Twice as natural," *Nature* 414 (2001): 255.

CHAPTER FIFTEEN: THE LURE OF COMPLEXITY

1. N. Bohr, "Light and life," *Nature* 131 (1933): 421–423, 457–459.

2. Ernst Peter Fischer and Carol Lipson, *Thinking About Science: Max Delbrück and the Origins of Molecular Biology* (New York: W. W. Norton, 1988), 131–156; see also D. J. McKaughan, "The influence of Niels Bohr on Max Delbrück: Revisiting the hopes inspired by 'Light and Life,'" *Isis* 96 (2005): 507–529.

3. The American cell biologist Marc Kirschner has described appeals to self-organization as "molecular vitalism"—that is, a means of accounting for characteristics of organisms that eighteenth- and nineteenth-century biologists could explain only by reference to a "vital force." See M. Kirschner, J. Gerhart, and T. Mitchison, "Molecular 'vitalism,'" *Cell* 100 (2000): 79–88.

4. See, for example, D. McShea, "Complexity and evolution: What everybody knows," *Biology and Philosophy* 6 (1991): 303–324.

5. R. Gallagher and T. Appenzeller, eds., "Beyond reductionism," *Science* 284 (1999): 79–109; L. Chong and L. B. Ray, eds., "Whole-istic biology," *Science* 295 (2002): 1661–1682; K. Ziemelis and L. Allen, eds., "Complex systems," *Nature* 410 (2001): 241–284; C. Surridge, ed., "Computational biology," *Nature* ("Insight" Supplement) 420 (2002): 205–251; and P. Aldhous, "Harvard's melting pot," *Nature* 416 (2002): 256–257.

6. The following list of books, ranging in order from specialized monographs to works intended for a general audience, may be recommended to interested readers: Jacques Ricard, *Biological Complexity and the Dynamics of Life Processes* (Amsterdam: Elsevier, 1999); Scott Camazine, Jean-Louis Deneubourg, Nigel R. Franks, James Sneyd, Guy Theraulaz, and Eric Bonabeau, *Self-Organization in Biological Systems* (Princeton: Princeton University Press, 2001); Uri Alon, *An Introduction to Systems Biology: Design Principles of Biological Circuits* (London: Chapman & Hall, 2006); Kunihiko Kaneko, *Life: An Introduction to Complex Systems Biology* (Berlin: Springer, 2006); Ricard Solé and Brian Goodwin, *Signs of Life: How Complexity Pervades Biology* (New York: Basic, 2000); Albert-Lázló Barabási, *Linked: The New Science of Networks* (Cambridge, Mass.: Perseus, 2002); Mark Buchanan,

Nexus: Small Worlds and the Groundbreaking Science of Networks (New York: W. W. Norton, 2002); and Mark Ward, *Beyond Chaos: The Underlying Theory Behind Life, the Universe, and Everything* (New York: Thomas Dunne/ St. Martin's, 2001).

7. Cohen and Stewart, "Where are the dolphins?" 1121.

8. For an attempt to provide a precise definition of emergence, see William C. Wimsatt, "Aggregativity: Reductive heuristics for finding emergence," *Philosophy of Science* 64 (1997): S372–S384.

9. M. A. O'Malley and J. Dupré, "Fundamental issues in systems biology," *BioEssays* 27 (2005): 1270–1276.

10. See, for example, Kauffman, *The Origin of Order,* 287–404, and Stuart A. Kauffman, *Investigations* (New York: Oxford University Press, 2000), 343–356.

11. Kauffman, *Investigations,* 72.

12. Ludwig von Bertalanffy, *General System Theory: Foundations, Development, Applications* (New York: George Braziller, 1968). That organisms are complex systems is a point made by Warren Weaver more than fifty years ago in "Science and complexity," *American Scientist* 36 (1948): 536–544.

13. H. Bénard, "Les tourbillons cellulaires dans une nappe liquide," *Revue générale des Sciences pures et appliquées* 11 (1900): 1261–1271, 1309–1328.

14. Joshua M. Epstein, *Nonlinear Dynamics, Mathematical Biology, and Social Science* (Reading, Mass.: Addison-Wesley, 1997), and *Generative Social Science: Studies in Agent-Based Computational Modeling* (Princeton: Princeton University Press, 2006).

CHAPTER SIXTEEN: THE THREE PILLARS OF LIFE

1. E. Mayr, "Cause and effect in biology," *Science* 134 (1961): 1501–1506.

2. As early as 1924, in *The Origin of Life,* Aleksandr Oparin argued that life is characterized not by any special property but by a definite and specific combination of fundamental properties. Even if what Oparin considered to be fundamental properties are not so regarded by biologists today, and despite his insistence on the absence of nonnatural characteristics in organisms, he clearly articulated the point that I wish to emphasize here: the secret of life lies in the conjunction of distinct properties, no one of which is by itself essential to life.

3. See Kamminga, "Historical perspective."

4. T. Surrey, F. Nédélec, S. Leibler, and E. Karsenti, "Physical properties determining self-organization of motors and microtubules," *Science* 292 (2001): 1167–1171.

5. Raymond Ruyer, *La genèse des formes vivantes* (Paris: Flammarion, 1958).

6. J. E. Lovelock, *Gaia: A New Look at Life on Earth* (Oxford: Oxford University Press, 1979).

7. M. G. Sacerdote and J. W. Szostak, "Semipermeable lipid bilayers exhibit diastereoselectivity favoring ribose," *Proc. Natl. Acad. Sci. USA* 102 (2005): 6004–6008.

8. D. E. Ingber, "The origins of cellular life," *BioEssays* 22 (2000): 1160–1170; see also T. Cavalier-Smith, "The origin of cells: A symbiosis between genes, catalysts, and membranes," *Cold Spring Harbor Symp. Quant. Biol.* 52 (1987): 805–824.

9. See, for example, S. Rasmussen, L. Chen, D. Deamer, D. C. Krakauer, N. H. Packard, P. F. Stadler, and M. A. Bedau, "Transitions from nonliving to living matter," *Science* 303 (2004): 963–965; R. V. Solé, S. Rasmussen, and M. Bedau, eds., "Theme issue: Towards the artificial cell," *Phil. Trans. R. Soc. B* 362 (2007): 1725–1925.

10. On Venter's recent progress in this direction see C. Lartigue, J. J. Glass, N. Alperovich, R. Pieper, and P. P. Parmar, et al., "Genome transplantation in bacteria: Changing one species to another," *Science* 317 (2007): 632–638.

11. Cho, Magnus, Caplan, et al., "Ethical considerations in synthesizing a minimal genome."

CONCLUSION

1. On robustness, see Andreas Wagner, *Robustness and Evolvability in Living Systems* (Princeton: Princeton University Press, 2005); the purposefulness of organisms is discussed at some length in Monod, *Chance and Necessity*, 5–22; on illness, see Georges Canguilhem, *The Normal and the Pathological*, trans. Carolyn R. Fawcett (New York: Zone, 1991), 203–226. This final property, which no inanimate object in the material world possesses, is probably linked to the ability of complex systems, in the face of a malfunction or alteration, to adopt another mode of functioning and interacting with the environment.

When I had nearly completed work on the French edition of this book I was surprised to see that Stephen Jay Gould, in his final work, had also employed a three-pillared logic in elaborating Darwinian theory. The ontological character of his supporting members is very different from that of mine, however. See *The Structure of Evolutionary Theory* (Cambridge, Mass.: Belknap Press of Harvard University Press, 2002).

2. Dominique Lecourt, "Théorie du vivant," in *Dictionnaire d'histoire et de la philosophie des sciences,* ed. Dominique Lecourt (Paris: Presses Universitaires de France, 1999), 989–992.

3. See the concluding chapter ("La notion de vie aujourd'hui") in Pichot, *Histoire de la notion de vie,* 936–954.

4. Frans de Waal, *Chimpanzee Politics: Power and Sex Among Apes,* rev. ed. (Baltimore: Johns Hopkins University Press, 1998).

5. E. G. Nisbet and N. H. Sleep, "The habitat and nature of early life," *Nature* 409 (2001): 1083.

6. For a more detailed description of the major transitions in evolutionary history, see John Maynard Smith and Eörs Szathmary, *The Origins of Life: From the Birth of Life to the Origin of Language* (Oxford: Oxford University Press, 1999).

Bibliography

Aldhous, P. "Harvard's melting pot." *Nature* 416 (2002): 256–257.

Aldiss, B. W. "Desperately seeking aliens." *Nature* 409 (2001): 1080–1082.

Alexander, J., and C. B. Bridge. "Some physiochemical aspects of life, mutation, and evolution." *Science* 70 (1929): 508–510.

Allwood, A. C., M. R. Walter, B. S. Kamber, C. P. Marshall, and I. W. Burch. "Stromatolites reef from the early Archaean era of Australia." *Nature* 441 (2006): 714–718.

Alon, Uri. *An Introduction to Systems Biology: Design Principles of Biological Circuits.* London: Chapman & Hall, 2006.

Ameisen, J.-C. "On the origin, evolution, and nature of programmed cell death: A timeline of four billion years." *Cell Death Differ.* 9 (2002): 367–393.

Andler, Daniel, Anne Fagot-Largeault, and Bertrand Saint-Sernin. *Philosophie des Sciences.* Paris: Gallimard, 2002.

Arrhenius, S. "The propagation of life in space." *Die Umschau* 7 (1903): 481.

Atkins, J. F., and R. Gesteland. "The 22nd amino acid." *Science* 296 (2002): 1409–1410.

Atlan, Henri. *La fin du "tout génétique"? Vers de nouveaux paradigmes en biologie.* Paris: INRA Éditions, 1999.

Bada, J. L. "State-of-the-art instruments for detecting extraterrestrial life." *Proc. Natl. Acad. Sci. USA* 98 (2001): 797–800.

Bada, J. L., and A. Lazcano. "Some like it hot, but not the first biomolecules." *Science* 296 (2002): 1982–1983.

Barabási, Albert-Lázló. *Linked: The New Science of Networks.* Cambridge, Mass.: Perseus, 2002.

Basalla, George. *Civilized Life in the Universe: Scientists on Intelligent Extraterrestrials.* Oxford: Oxford University Press, 2005.

Becquerel, P. [Untitled article.] *Bull. Soc. Astron.* 38 (1924): 393.

Bénard, H. "Les tourbillons cellulaires dans une nappe liquide." *Revue générale des Sciences pures et appliquées* 11 (1900): 1261–1271, 1309–1328.

Benner, S. A., A. D. Ellington, and A. Tauer. "Modern metabolism as a palimpsest of the RNA world." *Proc. Natl. Acad. Sci. USA* 86 (1989): 7054–7058.

Bernal, J. D., ed. *Origin of Life.* London: Weidenfeld and Nicolson, 1967.

Bernard, Claude. *Phenomena of Life Common to Animals and to Plants.* Trans. R. P. Cook and M. A. Cook. Dundee: Cook & Cook, 1974.

Bertalanffy, Ludwig von. *General System Theory: Foundations, Development, Applications.* New York: George Braziller, 1968.

Bibring, J.-P., Y. Langevin, J. F. Mustard, et al. "Global mineralogical and aqueous Mars history derived from OMEGA/Mars Express data." *Science* 312 (2006): 400–404.

Bichat, Xavier. *Recherches physiologiques sur la vie et la mort.* Ed. André Pichot. Paris: Flammarion, 1994.

Biemann, K., J. Oro, P. Toulmin III, et al. "Search for organic and volatile inorganic compounds in two surface samples from the Chryse Planitia Region of Mars." *Science* 194 (1976): 72–76.

Bohr, N. "Light and life." *Nature* 131 (1933): 421–423, 457–459.

Borges, Jorge Luis. *The Book of Imaginary Beings.* Trans. Norman Thomas di Giovanni. New York: Dutton, 1969.

Brasier, M. D., O. R. Green, A. P. Jephcoat, et al. "Questioning the evidence for Earth's oldest fossils." *Nature* 416 (2002): 76–81.

Buchanan, Mark. *Nexus: Small Worlds and the Groundbreaking Science of Networks.* New York: W. W. Norton, 2002.

Bud, Robert. *The Uses of Life: A History of Biotechnology.* Cambridge: Cambridge University Press, 1993.

Burt, Austin, and Robert Trivers. *Genes in Conflict: The Biology of Selfish Genetic Elements.* Cambridge, Mass.: Belknap Press of Harvard University Press, 2006.

Bushman, Frederic. *Lateral DNA Transfer: Mechanisms and Consequences.* Cold Spring Harbor, N.Y.: Cold Spring Harbor Laboratory Press, 2002.

Cairns-Smith, A. G. "The origin of life and the nature of the primitive gene." *J. Theoret. Biol.* 10 (1965): 53–88.

Camazine, Scott, Jean-Louis Deneubourg, Nigel R. Franks, James Sneyd, Guy Theraulaz, and Eric Bonabeau. *Self-Organization in Biological Systems.* Princeton: Princeton University Press, 2001.

Canguilhem, Georges. *La connaissance de la vie.* Paris: Vrin, 1975.

———. "Vie." In *Encyclopaedia Universalis,* 32 vols. Paris: Éditions de l'Encyclopaedia Universalis, 1989, 24:546–553.

———. *A Vital Rationalist: Selected Writings from Georges Canguilhem.* Ed. François Delaporte, trans. Arthur Goldhammer. New York: Zone, 1994.

Cardillo, M., G. M. Mace, K. E. Jones, et al. "Multiple causes of high extinction risk in large mammal species." *Science* 309 (2005): 1239–1241.

Carroll, S. B. "Chance and necessity: The evolution of morphological complexity and diversity." *Nature* 409 (2001): 1102–1109.

Cash, W. "Detection of Earth-like planets around nearby stars using a petal-shaped occulter." *Nature* 442 (2006): 51–53.

Cavalier-Smith, T. "The origin of cells: A symbiosis between genes, catalysts, and membranes." *Cold Spring Harbor Symp. Quant. Biol.* 52 (1987): 805–824.

Cech, T. R. "The ribosome is a ribozyme." *Science* 289 (2000): 878–879.

Cello, J., A. V. Paul, and E. Wimmer. "Chemical synthesis of poliovirus cDNA: generation of infectious virus in the absence of natural template." *Science* 297 (2002): 1016–1018.

Chen, I. A., R. W. Roberts, and Jack W. Szostak. "The emergence of competition between model protocells." *Science* 305 (2004): 1474–1476.

Cho, M. K., D. Magnus, A. L. Caplan, et al. "Ethical considerations in synthesizing a minimal genome." *Science* 286 (1999): 2087–2090.

Chong, L., and L. B. Ray, eds. "Whole-istic biology," *Science* 295 (2002): 1661–1682.

Christensen, P. R. "Formation of recent Martian gullies through melting of extensive water-rich snow deposits." *Nature* 422 (2003): 45–48.

Chyba, C. F. "Energy for microbial life on Europa." *Nature* 403 (2000): 381–382.

Chyba, C. F., and C. B. Phillips. "Possible ecosystems and the search for life on Europa." *Proc. Natl. Acad. Sci. USA* 98 (2001): 801–804.

Cody, G. D. "Transition metal sulfides and the origin of metabolism." *Ann. Rev. Earth planetary Sci.* 32 (2004): 569–599.

Cohen, Jack S., and Ian Stewart. "Where are the dolphins?" *Nature* 409 (2001): 1119–1122.

———. *Evolving the Alien: The Science of Extraterrestrial Life.* London: Ebury Press/John Wiley, 2002.

Creager, Angela N. H. *The Life of a Virus: Tobacco Mosaic Virus as an Experimental Model.* Chicago: University of Chicago Press, 2002.

Crick, F. "The origin of the genetic code." *J. Mol. Biol.* 38 (1968): 367–379.

Crick, F., and L. E. Orgel. "Directed panspermia." *Icarus* 19 (1973): 341–346.

Cunchillos, C., and G. Lecointre. "Evolution of amino acid metabolism inferred through cladistic analysis." *J. Biol. Chem.* 278 (2003): 47960–47970.

———. "Integrating the universal metabolism into a phylogenetic analysis." *Mol. Biol. Evol.* 22 (2005): 1–11.

Cunningham, B. "The reemergence of 'emergence.'" *Philosophy of Science* 68 (2001): S62–S75.

Danchin, Antoine. *The Delphic Boat: What Genomes Tell Us.* Trans. Alison Quayle. Cambridge, Mass.: Harvard University Press, 2002.

Darwin, Charles. *The Origin of Species.* London: J. M. Dent, 1972.

David, Patrice, and Sarah Samadi. *La théorie de l'évolution: Une logique pour la biologie.* Paris: Flammarion, 2000.

Davis, Rowland H. *The Microbial Models of Molecular Biology: From Genes to Genomes.* Oxford: Oxford University Press, 2003.

Dawkins, Richard. *The Selfish Gene.* Oxford: Oxford University Press, 1976; 30th anniversary edition, 2006.

Deamer, David W., and Gail R. Fleischaker, eds. *Origins of Life: The Central Concepts.* Boston: Jones and Bartlett, 1994.

Delbrück, M. "Preliminary write-up on the topic 'Riddle of Life.'" In "A physicist's renewed look at biology: Twenty years later." *Science* 168 (1970): 1314–1315.

Denamur, E., G. Lecointre, P. Darlu, et al. "Evolutionary implications of the frequent horizontal transfer of mismatch repair genes." *Cell* 103 (2000): 711–721.

Dick, S. J. "NASA and the search for life in the universe." *Endeavour* 30 (2006): 71–75.

Dick, Steven J., and James L. Strick. *The Living Universe: NASA and the Development of Astrobiology.* New Brunswick, N.J.: Rutgers University Press, 2004.

Dietrich, D. E., and J. T. Perron. "The search for a topographic signature of life." *Nature* 439 (2006): 411–418.

Döring, V., H. D. Mootz, L. A. Nangle, et al. "Enlarging the amino acid set of *Escherichia coli* by infiltration of the valine coding pathway." *Science* 292 (2001): 501–504.

Duchesneau, François. *Philosophie de la biologie.* Paris: Presses Universitaires de France, 1997.

Dupuy, Jean-Pierre. *On the Origins of Cognitive Science: The Mechanization of the Mind.* Trans. M. B. DeBevoise. Cambridge, Mass.: The MIT Press, 2008.

Dürr, Hans Peter, Fritz-Albert Popp, and Wolfram Schommers, eds. *What Is Life? Scientific Approaches and Philosophical Positions.* River Edge, N.J.: World Scientific, 2002.

Duve, Christian de. *Vital Dust: Life as a Cosmic Imperative.* New York: Basic, 1995.

———. *Life Evolving: Molecules, Mind, and Meaning.* Oxford: Oxford University Press, 2002.

———. *Singularities: Landmarks on the Pathways of Life.* Cambridge: Cambridge University Press, 2005.

Dyson, Freeman. *Origins of Life.* 2nd revised edition. Cambridge: Cambridge University Press, 1999.

[Editorial.] "Meanings of 'life.'" *Nature* 447 (2007): 1031–1032.

Ehrenfreund, P., D. P. Glavin, O. Botta, et al. "Extraterrestrial amino acids in Orgueil and Ivuna: Tracing the parent body of CI type carbonaceous chondrites." *Proc. Natl. Acad. Sci. USA* 98 (2001): 2138–2141.

Eigen, M., C. K. Biebricher, M. Gebinoga, and W. C. Gardiner. "The hypercycle: Coupling of RNA and protein biosynthesis in the infection cycle of an RNA bacteriophage." *Biochemistry* 30 (1991): 11005–11018.

Eigen, M., W. Gardiner, P. Schuster, and R. Winkler-Oswatitsch. "The origin of genetic information." *Sci. Am.* 244 (1981): 78–94.

Eldredge, Niles. *Reinventing Darwin: The Great Debate at the High Table of Evolutionary Theory.* New York: Wiley, 1995.

Ellington, A. D., and J. W. Szostak. "*In vitro* selection of RNA molecules that bind specific ligands." *Nature* 346 (1990): 818–822.

Embley, T. M., and W. Martin. "Eukaryotic evolution, changes and challenges." *Nature* 440 (2006): 623–630.

Engelberg-Kulka, H., and G. Glaser. "Addiction molecules and programmed cell death and anti-death in bacterial cultures." *Annu. Rev. Microbiol.* 53 (1999): 43–70.

Engels, Friedrich. *Anti-Dühring: Herr Eugen Dühring's Revolution in Science.* Moscow: Progress, 1947.

Epstein, Joshua M. *Nonlinear Dynamics, Mathematical Biology, and Social Science.* Reading, Mass.: Addison-Wesley, 1997.

———. *Generative Social Science: Studies in Agent-Based Computational Modeling.* Princeton: Princeton University Press, 2006.

Erwin, Douglas H. *Extinction: How Life on Earth Nearly Ended 250 Million Years Ago.* Princeton: Princeton University Press, 2006.

Falkowski, P. G., M. E. Katz, A. J. Milligan, et al. "The rise of oxygen over the past 205 million years and the evolution of large placental mammals." *Science* 309 (2005): 2202–2204.

Fedo, C. M., and M. J. Whitehouse. "Metasomatic origin of quartz-pyroxene rock, Akilia, Greenland, and implications for Earth's earliest life." *Science* 296 (2002): 1448–1452.

Fernández Ostolaza, Julio, and Álvaro Moreno Bergareche. *Vida artificial.* Madrid: Eudema, 1992.

Fischer, Ernst Peter, and Carol Lipson. *Thinking About Science: Max Delbrück and the Origins of Molecular Biology.* New York: W. W. Norton, 1988.

Fleischaker, G. R. "Origins of life: An operational definition." *Orig. Life Evol. Biosph.* 20 (1990): 127–137.

Forterre, P. "L'origine du génome." *La Recherche* 336 (2000): 34–39.

———. "The origin of viruses and their possible roles in major evolutionary transitions." *Virus Research* 117 (2006): 5–16.

Foucault, Michel. *The Order of Things: An Archaeology of the Human Sciences.* [No translator indicated.] New York: Pantheon, 1971.

Fox, S. W. "Spontaneous generation, the origin of life, and self-assembly." *Curr. Mod. Biol.* 2 (1968): 235–240.

Fraser, C. M., J. D. Gocayne, O. White, et al. "The minimal gene complement of *Mycoplasma genitalium.*" *Science* 270 (1995): 397–403.

Freeland, S. J., R. D. Knight, and L. F. Landweber. "Do proteins predate DNA?" *Science* 286 (1999): 690–692.

Fry, Ris. *The Emergence of Life on Earth: A Historical and Scientific Overview.* New Brunswick, N.J.: Rutgers University Press, 2000.

Fuqua, C., S. C. Winans, and E. P. Greenberg. "Census and concensus in bacterial ecosystems: The LuxR-LuxI family of quorum-sensing transcriptional regulators." *Annu. Rev. Microbiol.* 50 (1996): 727–751.

Furnes, H., N. R. Banerjee, K. Muehlenbachs, et al. "Early life recorded in Archaean pillow lavas." *Science* 304 (2004): 578–581.

Fusz, S., A. Elsenführ, S. G. Srivatsan, A. Heckel, and M. Famulok. "A ribozyme for the aldol reaction." *Chemistry and Biology* 12 (2005): 941–950.

Gallagher, R., and T. Appenzeller, eds. "Beyond reductionism." *Science* 284 (1999): 79–109.

Galtier, N., N. Tourasse, and M. Gouy. "A nonhyperthermophilic common ancestor to extant life-forms." *Science* 283 (1999): 220–221.

Gánti, Tibor. *The Principles of Life.* Oxford: Oxford University Press, 2003; originally published in Budapest in 1971.

García-Ruiz, J. M., S. T. Hyde, A. M. Carnerup, et al. "Self-assembled silica-carbonate structures and detection of ancient microfossils." *Science* 302 (2003): 1194–1197.

Gaucher, E. A., J. M. Thomson, M. F. Burgan, and S. A. Benner. "Inferring the palaeoenvironment of ancient bacteria on the basis of resurrected proteins." *Nature* 425 (2003): 285–288.

Gavaghan, H. "ESA embraces astrobiology." *Science* 292 (2001): 1626–1627.

Gayon, Jean. *Darwinism's Struggle for Survival: Heredity and the Hypothesis of Natural Selection.* Trans. Matthew Cobb. Cambridge: Cambridge University Press, 1998.

———. "From Darwin to Today in Evolutionary Biology." In *The Cambridge Companion to Darwin,* ed. Hodge and Radick, 240 – 264.

Gil, R., F. J. Silva, J. Pereto, and A. Moya. "Determination of the core of a minimal bacterial gene set." *Microbiol. Mol. Biol. Rev.* 68 (2004): 518 – 537.

Gilbert, W. "The RNA world." *Nature* 319 (1986): 319, 618.

Gilson, P. R., V. Su, C. H. Slamovits, et al. "Complete nucleotide sequence of the chlorarachniophyte nucleomorph: Nature's smallest nucleus." *Proc. Natl. Acad. Sci. USA* 103 (2006): 9566 – 9571.

Gould, Stephen Jay. *Wonderful Life: The Burgess Shale and the Nature of History.* New York: W. W. Norton, 1989.

———. "Darwinian Fundamentalism." *New York Review of Books,* 12 June 1997.

———. *The Structure of Evolutionary Theory.* Cambridge, Mass.: Belknap Press/Harvard University Press, 2002.

Green, R. G. "On the nature of filterable viruses." *Science* 82 (1935): 443 – 445.

Grinspoon, David. *Lonely Planets: The Natural Philosophy of Alien Life.* New York: HarperCollins, 2004.

Haldane, J. B. S. "The Origin of Life." In *The Rationalist Annual,* ed. Charles A. Watts. London: Watts, 1929.

———. "The origins of life." *New Biology* 16 (1954): 12 – 27.

Hanczyc, M. M., S. M. Fujikawa, and J. W. Szostak. "Experimental models of primitive cellular compartments: Encapsulation, growth, and division." *Science* 302 (2003): 618 – 622.

Hazen, Robert M. *Genesis.* Washington, D.C.: Joseph Henry Press, 2005.

d'Herelle, Félix. *Les défenses de l'organisme.* Paris: Flammarion, 1923.

Hey, Jody. *Genes, Categories, and Species: The Evolutionary and Cognitive Causes of the Species Problem.* New York: Oxford University Press, 2001.

Hirao, I., T. Ohtsuki, T. Fujiwara, et al. "An unnatural base pair for incorporating amino acid analogs into proteins." *Nature Biotechnology* 20 (2002): 177 – 182.

Hodge, Jonathan, and Gregory Radick, eds. *The Cambridge Companion to Darwin.* Cambridge: Cambridge University Press, 2003.

Holton, Gerald. *The Scientific Imagination: Case Studies.* Cambridge: Cambridge University Press, 1978.

d'Hondt, S., Jorgensen, B. B., Miller, D. J., et al. "Distributions of microbial activities in deep subseafloor sediments." *Science* 306 (2004): 2216 – 2221.

Horowitz, N. H., R. E. Cameron, and J. S. Hubbard. "Microbiology of the dry valleys of Antarctica." *Science* 176 (1972): 242–245.

Hoyle, Fred, and Chandra Wickramasinghe. *Our Place in the Cosmos: The Unfinished Revolution?* London: Phoenix, 1996.

Huber, C., and G. Wächtershäuser. "α-Hydroxy and α-amino acids under possible Hadean, volcanic origin of life conditions." *Science* 314 (2006): 630–632.

Huber, H., M. J. Hohn, R. Rachel, et al. "A new phylum of archaea represented by a nanosized hyperthermophilic symbiont." *Nature* 417 (2002): 63–67.

Hutchison, C. A. III, S. N. Peterson, S. R. Gill, et al. "Global transposon mutagenesis and a minimal mycoplasma genome." *Science* 286 (1999): 2165–2169.

Huynen, M. "Constructing a minimal genome." *Trends in Genetics* 16 (2000): 116.

Ingber, D. E. "The origin of cellular life." *BioEssays* 22 (2000): 1160–1170.

Irwin, R. P. III, T. A. Maxwell, A. D. Howard, et al. "A large paleolake basin at the head of Ma'adim Vallis, Mars." *Science* 296 (2002): 2209–2212.

Israël, G., C. Szopa, F. Raulin, et al. "Complex organic matter in Titan's atmospheric aerosols from *in situ* pyrolysis and analysis." *Nature* 438 (2005): 796–799.

Jablonka, Eva, and Marion J. Lamb. *Evolution in Four Dimensions: Genetic, Epigenetic, Behavioral, and Symbolic Variation in the History of Life.* Cambridge, Mass.: The MIT Press, 2005.

Jacob, François. "The Leeuwenhoek Lecture, 1977: Mouse teratocarcinoma and mouse embryo." *Proc. Roy. Soc. London Ser. B.*, 201 (1978): 249–270.

———. *The Possible and the Actual.* New York: Pantheon, 1982.

———. *The Logic of Life: A History of Heredity.* Trans. Betty E. Spillmann. Princeton: Princeton University Press, 1993.

Jacob, F., and J. Monod. "Genetic regulatory mechanisms in the synthesis of proteins." *J. Mol. Biol.* 3 (1961): 318–356.

Johnston, W. K., P. J. Unrau, M. S. Lawrence, et al. "RNA-catalyzed RNA polymerization: Accurate and general RNA-templated primer extension." *Science* 292 (2001): 1319–1325.

Joint, I., J. A. Downie, and P. Williams, eds. "Bacterial conversations: Talking, listening, and eavesdropping." Special issue, *Phil. Trans. R. Soc. B*, 29 July 2007.

Joyce, G. F. "The antiquity of RNA-based evolution." *Nature* 418 (2002): 214–221.

Kahane, Ernest. *La vie n'existe pas.* Paris: Éditions Rationalistes, 1962.

Kajander, E. O., and N. Ciftcioglu. "Nanobacteria: An alternative mechanism for pathogenic intra- and extracellular calcification and stone formation." *Proc. Natl. Acad. Sci. USA* 95 (1998): 8274–8279.

Kamminga, H. "Historical perspective: The problem of the origin of life in the context of developments in biology." *Orig. Life Evol. Biosph.* 18 (1988): 1–11.

Kaneko, Kunihiko. *Life: An Introduction to Complex Systems Biology.* Berlin: Springer, 2006.

Kang, K.-I., and M. Morange. "Succès et limites de l'étude moléculaire de la mort cellulaire programmée." *Annales d'histoire et de philosophie du vivant* 4 (2001): 159–175.

Kauffman, Stuart A. *The Origin of Order: Self-Organization and Selection in Evolution.* New York: Oxford University Press, 1993.

———. *At Home in the Universe: The Search for the Laws of Self-Organization and Complexity.* New York: Oxford University Press, 1995.

———. *Investigations.* New York: Oxford University Press, 2000.

Kay, L. E. "W. M. Stanley's crystallization of the tobacco mosaic virus, 1930–1940." *Isis* 77 (1986): 450–472.

Keller, Evelyn Fox. *The Century of the Gene.* Cambridge, Mass.: Harvard University Press, 2000.

———. *Making Sense of Life: Explaining Biological Development with Models, Metaphors, and Machines.* Cambridge, Mass.: Harvard University Press, 2002.

Kerr, R. A. "Putting Martian science to the test." *Science* 301 (2003): 1832–1834.

Kirschner, Marc W., and John C. Gerhart. *The Plausibility of Life: Resolving Darwin's Dilemma.* New Haven: Yale University Press, 2005.

Kirschner, M., J. Gerhart, and T. Mitchison. "Molecular 'vitalism,'" *Cell* 100 (2000): 79–88.

Kivelson, M. G., K. K. Khurana, C. T. Russell, et al. "Galileo magnetometer measurements: A stronger case for a subsurface ocean at Europa." *Science* 289 (2000): 1340–1343.

Klein, H. P., N. H. Horowitz, G. V. Levin, et al. "The Viking biological investigation: Preliminary results." *Science* 194 (1976): 99–105.

Klussmann, M., H. Iwamura, S. P. Mathew, et al. "Thermodynamic control of asymmetric amplification in amino acid catalysis." *Nature* 441 (2006): 621–623.

Knauth, L. P., D. M. Burt, and K. H. Wohletz. "Impact origin of sediments at the Opportunity landing site on Mars." *Nature* 438 (2005): 1123–1128.

Knoll, A. H., and S. B. Carroll. "Early animal evolution: Emerging views from comparative biology and geology." *Science* 284 (1999): 2129–2137.

Kolter, R., and E. P. Greenberg. "The superficial life of microbes." *Nature* 441 (2006): 300–302.

Koshland, D. E., Jr. "The seven pillars of life." *Science* 295 (2002): 2215–2216.

Kupiec, Jean-Jacques, and Pierre Sonigo. *Ni Dieu ni gène: Pour une autre théorie de l'hérédité.* Paris: Seuil, 2000.

Kurland, C. G., L. J. Collins, and D. Penny. "Genomics and the irreducible nature of eukaryotic cells." *Science* 312 (2006): 1011–1014.

Lamarck, Jean-Baptiste. *Zoological Philosophy: An Exposition with Regard to the Natural History of Animals.* Trans. Hugh Elliot. London: Hafner, 1963.

Langton, Christopher G., ed. *Artificial Life.* Vol. 6 of *Santa Fe Studies in the Sciences of Complexity.* Reading, Mass.: Addison-Wesley, 1989.

Lartigue, C., J. J. Glass, N. Alperovich, R. Pieper, and P. P. Parmar, et al. "Genome transplantation in bacteria: Changing one species to another." *Science* 317 (2007): 632–638.

Lawrence, M. S., and D. P. Bartel. "New ligase-derived RNA polymerase ribozymes." *RNA* 11 (2005): 1173–1180.

Lecourt, Dominique, ed. *Dictionnaire d'histoire et de la philosophie des sciences.* Paris: Presses Universitaires de France, 1999.

Lederberg, J. "Exobiology: Approaches to life beyond the Earth." *Science* 132 (1960): 393–400.

Lenski, R. E. "Twice as natural." *Nature* 414 (2001): 255.

"Les frontières du vivant." Special issue. *La Recherche* 317 (1999).

Lewontin, R. C. "The units of selection." *Annu. Rev. Ecol. System* 1 (1970): 1–18.

Libera, Alain de. *La querelle des universaux: De Platon à la fin du Moyen Âge.* Paris: Le Seuil, 1996.

Lineweaver, C. H., Y. Fenner, and B. K. Gibson. "The galactic habitable zone and the age distribution of complex life in the Milky Way." *Science* 303 (2004): 59–62.

Lissauer, J. J. "Extrasolar planets." *Nature* 419 (2002): 355–358.

Loewenstein, Werner R. *The Touchstone of Life: Molecular Information, Cell Communication, and the Foundations of Life.* New York: Oxford University Press, 1999.

Lorenz, Ralph, and Jacqueline Mitton. *Lifting Titan's Veil: Exploring the Giant Moon of Saturn.* Cambridge: Cambridge University Press, 2002.

Lovelock, J. E. "A physical basis for life detection experiments." *Nature* 207 (1965): 568–570.

———. *Gaia: A New Look at Life on Earth.* Oxford: Oxford University Press, 1979.

Lovis, C., M. Mayor, F. Pepe, et al. "An extrasolar planetary system with three Neptune-mass planets." *Nature* 441 (2006): 305–309.

Luisi, Pier Luigi. "About various definitions of life." *Orig. Life Evol. Biosph.* 28 (1998): 613–622.

———. *The Emergence of Life: From Chemical Origins to Synthetic Biology.* Cambridge: Cambridge University Press, 2006.

Lwoff, André. *L'évolution physiologique: Étude des pertes de fonction chez les microorganismes.* Paris: Hermann, 1944.

———. "The concept of virus." *J. Gen. Microbiol.* 17 (1957): 239–253.

McCollom, T. M., and B. M. Hynek. "A volcanic environment for bedrock diagenesis at Meridiani Planum on Mars." *Nature* 438 (2005): 1129–1131.

McCord, T. B., G. B. Hansen, and C. A. Hibbitts. "Hydrated salt minerals on Ganymede's surface: Evidence of an ocean below." *Science* 292 (2001): 1523–1525.

McGinness, K. E., and G. F. Joyce. "In search of an RNA replicase ribozyme." *Chem. Biol.* 10 (2003): 5–14.

McKaughan, D. J. "The influence of Niels Bohr on Max Delbrück: Revisiting the hopes inspired by 'Light and Life.'" *Isis* 96 (2005): 507–529.

McKay, D. S., E. K. Gibson, Jr., K. L. Thomas-Keprta, et al. "Search for past life on Mars: Possible relic biogenic activity in Martian meteorite ALH84001." *Science* 273 (1996): 924–930.

McKeon, Richard, ed. *The Basic Works of Aristotle.* New York: Random House, 1941.

McShea, D. "Complexity and evolution: What everybody knows." *Biology and Philosophy* 6 (1991): 303–324.

Malin, M. C., and K. S. Edgett. "Evidence for recent groundwater seepage and surface runoff on Mars." *Science* 288 (2000): 2330–2335.

Malin, M. C., K. S. Edgett, L. V. Posiolova, S. M. McColley, and E. Z. N. Dobrea. "Evidence for persistent flow and aqueous sedimentation on early Mars." *Science* 302 (2003): 1931–1934.

———. "Present-day impact cratering rate and contemporary gully activity on Mars." *Science* 314 (2006): 1573–1577.

Margulis, Lynn, and Dorion Sagan. *What Is Life?* Berkeley: University of California Press, 1995.

Mayr, E. "Cause and effect in biology." *Science* 134 (1961): 1501–1506.

Merleau-Ponty, Maurice. *Le visible et l'invisible.* Paris: Gallimard, 1964.

Miller, S. L. "Production of amino acids under possible primitive Earth conditions." *Science* 117 (1953): 528.

Monod, Jacques. *Chance and Necessity: An Essay on the Natural Philosophy of Modern Biology.* Trans. Austryn Wainhouse. New York: Knopf, 1971.

Moore, Janice. *Parasites and the Behavior of Animals.* Oxford: Oxford University Press, 2002.

Morange, Michel. *A History of Molecular Biology.* Trans. Matthew Cobb. Cambridge, Mass.: Harvard University Press, 1998.

———. "Gene function." *C. R. Acad. Sci. Paris, Sciences de la vie* 323 (2000): 1147–1153.

———. *The Misunderstood Gene.* Trans. Matthew Cobb. Cambridge, Mass.: Harvard University Press, 2001.

———. "Fifty years ago: The beginnings of exobiology." *J. Biosci.* 32 (2007): 1083–1087.

Morowitz, Harold J. *Beginnings of Cellular Life: Metabolism Recapitulates Biogenesis.* New Haven: Yale University Press, 1992.

Morris, Simon Conway. *Life's Solution: Inevitable Humans in a Lonely Universe.* New York: Cambridge University Press, 2003.

Muller, H. J. "Variation due to change in the individual gene," *The American Naturalist* 56 (1922): 32–50.

———. "The Gene as the Basis of Life." In *Proceedings of the International Congress of Plant Science, Ithaca, New York, August 16–23, 1926,* ed. B. M. Duggar. 2 vols. Menosha, Wis.: G. Banta, 1929, 1:897–921.

Mushegian, A. R., and E. V. Koonin. "A minimal gene set for cellular life derived by comparison of complete bacterial genomes." *Proc. Natl. Acad. Sci. USA* 93 (1996): 10268–10273.

Nadis, S. "Spoof Nobels take researchers for a ride." *Nature* 425 (2003): 550.

National Research Council, Committee on the Limits of Organic Life in Planetary Systems et al. *The Limits of Organic Life in Planetary Systems.* Washington, D.C.: National Academies Press, 2007.

Nisbet, E. G., and N. H. Sleep. "The habitat and nature of early life." *Nature* 409 (2001): 1083–1091.

Odling-Smee, John, Kevin Laland, and Marcus Feldman. *Niche Construction: The Neglected Process in Evolution.* Princeton: Princeton University Press, 2003.

Oparin, A. I. *The Origin of Life.* Trans. Sergius Morgulis. New York: Macmillan, 1938.

Orgel, L. E. "Evolution of the genetic apparatus." *J. Mol. Biol.* 38 (1968): 381–393.

———. "A simpler nucleic acid." *Science* 290 (2000): 1306–1307.

———. "Prebiotic chemistry and the origin of the RNA world." *Crit. Rev. Biochem. Mol. Biol.* 39 (2004): 99–123.

Oro, J. "Comets and the formation of biochemical compounds on the primitive Earth." *Nature* 190 (1961): 389–390.

Oster, G. "Darwin's motors." *Nature* 417 (2002): 25.

Ostwald, Wolfgang. *Introduction to Theoretical and Applied Colloid Chemistry: The World of Neglected Dimensions.* Trans. Martin H. Fischer. New York: John Wiley and Sons, 1917.

Oyama, Susan, Paul Griffiths, and Russell D. Gray. *Cycles of Contingency: Developmental Systems and Evolution.* Cambridge, Mass.: The MIT Press, 2001.

Pace, N. R. "The universal nature of biochemistry." *Proc. Natl. Acad. Sci. USA* 98 (2001): 805–808.

Pace, N. R., and T. L. Marsh. "RNA catalysis and the origin of life." *Orig. Life* 16 (1985): 97–116.

Pal, C., B. Papp, M. J. Lercher, et al. "Chance and necessity in the evolution of minimal metabolic networks." *Nature* 440 (2006): 667–670.

Pérez-Brocal, V., R. Gil, S. Ramos, A. Lamelas, M. Postigo, et al. "A small microbial genome: The end of a long symbiotic relationship?" *Science* 314 (2006): 312–313.

Perrett, J. "Biochemistry and bacteria." *New Biology* 12 (1952): 68–69.

Pezo, V., D. Metzgar, T. L. Hendrickson, et al. "Artificially ambiguous genetic code confers growth yield advantage." *Proc. Natl. Acad. Sci. USA* 101 (2004): 8593–8597.

Pichot, André. *Histoire de la notion de vie.* Paris: Gallimard, 1993.

Pirie, N. W. "The nature and development of life." *Modern Quarterly* 3 (1948): 82–93.

Podolsky, S. "The role of the virus in origin-of-life theorizing." *J. Hist. Biol.* 29 (1996): 79–126.

Popa, Radu. *Between Necessity and Probability: Searching for the Definition and Origin of Life.* Berlin: Springer, 2004.

Raoult, D., S. Audic, C. Robert, et al. "The 1.2-megabase genome sequence of mimivirus." *Science* 306 (2004): 1344–1350.

Rasmussen, S., L. Chen, D. Deamer, D. C. Krakauer, N. H. Packard, P. F. Stadler, and M. A. Bedau. "Transitions from nonliving to living matter." *Science* 303 (2004): 963–965.

Raulin-Cerceau, Florence, Pierre Léna, and Jean Schneider, eds. *Sur les traces du vivant: De la Terre aux Étoiles.* Paris: Le Pommier, 2002.

Ravin, A. W. "The gene as catalyst; the gene as organism." *Stud. Hist. Biol.* 1 (1977): 1–45.

Reichhardt, T. "Two telescopes join hunt for ET." *Nature* 440 (2006): 853.

Ricard, Jacques. *Biological Complexity and the Dynamics of Life Processes.* Amsterdam: Elsevier, 1999.

Rivera, M. C., and J. A. Lake. "The ring of life provides evidence for a genome fusion origin of eukaryotes." *Nature* 431 (2004): 152–155.

Robertson, D. L., and G. F. Joyce. "Selection *in vitro* of an RNA enzyme that specifically cleaves single-stranded DNA." *Nature* 344 (1990): 467–468.

Ronneberg, T. A., L. F. Landweber, and S. J. Freeland. "Testing a biosynthetic theory of the genetic code: Fact or artefact?" *Proc. Natl. Acad. Sci. USA* 97 (2000): 13690–13695.

Rothschild, L. J., and R. L. Mancinelli. "Life in extreme environments." *Nature* 409 (2001): 1092–1100.

Ruiz-Mirazo, K., J. Pereto, and A. Moreno. "A universal definition of life: Autonomy and open-ended evolution." *Orig. Life Evol. Biosph.* 34 (2004): 323–346.

Ruse, Michael. *Philosophy of Biology Today.* Albany: State University of New York Press, 1988.

Ruyer, Raymond. *La genèse des formes vivantes.* Paris: Flammarion, 1958.

Sacerdote, M. G., and J. W. Szostak. "Semipermeable lipid bilayers exhibit diastereoselectivity favoring ribose." *Proc. Natl. Acad. Sci. USA* 102 (2005): 6004–6008.

Sagan, C., W. R. Thompson, R. Carlson, et al. "A search for life on Earth from the Galileo spacecraft." *Nature* 365 (1993): 715–721.

Sapp, Jann. *Evolution by Association: A History of Symbiosis.* Oxford: Oxford University Press, 1994.

Sawyer, Kathy. *The Rock from Mars: A Detective Story on Two Planets.* New York: Random House, 2006.

Schilling, G. "Habitable, but not much like home." *Science* 316 (2007): 528.

Schmid-Hempel, P. "The awesome power of natural selection." *Nature* 417 (2002): 592.

Schopf, J. William. "Microfossils of the Early Archaean Apex chert: New evidence of the antiquity of life." *Science* 260 (1993): 640–646.

———. *Cradle of Life: The Discovery of Earth's Earliest Fossils.* Princeton: Princeton University Press, 1999.

———, ed. *Life's Origin: The Beginning of Biological Evolution.* Berkeley: University of California Press, 2002.

Schopf, J. W., A. B. Kudryavtsev, D. G. Agresti, et al. "Laser-Raman imagery of Earth's earliest fossils." *Nature* 416 (2002): 73–76.

Schrödinger, Erwin. *What Is Life? The Physical Aspect of the Living Cell.* Cambridge: Cambridge University Press, 1944; reprinted with foreword by Roger Penrose, 1992.

Segré, D., D. Ben-Eli, D. W. Deamer, and D. Lancet. "The lipid world." *Orig. Life Evol. Biosph.* 31 (2001): 119–145.
Service, R. F. "Creation's seventh day." *Science* 289 (2000): 232–235.
Shapiro, Robert. *Origins: A Skeptic's Guide to the Creation of Life on Earth.* New York: Summit, 1986.
———. "Prebiotic ribose synthesis: A critical analysis." *Orig. Life Evol. Biosph.* 18 (1988): 71–85.
Shaw, K. L. "Do we need species concepts?" *Science* 295 (2002): 1238–1239.
Shelley, Mary. *Frankenstein, or, The Modern Prometheus.* Ed. M. K. Joseph. New York: Oxford University Press, 1998.
Shigenobu, S., H. Watanabe, M. Hatori, et al. "Genome sequence of the endocellular bacterial symbiont of aphids *Buchnera* sp. APS." *Nature* 407 (2000): 81–86.
Shostak, Stanley. *Death of Life: The Legacy of Molecular Biology.* London: Macmillan, 1998.
Siegel, S. M., K. Roberts, H. Nathan, and O. Daly. "Living relative of the microfossil *Kakabekia.*" *Science* 158 (1967): 1231–1234.
Simpson, G. G. "The nonprevalence of humanoids." *Science* 143 (1964): 769–775.
Smith, John Maynard, and Eörs Szathmary. *The Origins of Life: From the Birth of Life to the Origin of Language.* Oxford: Oxford University Press, 1999.
Snow, C. P. *The Two Cultures and the Scientific Revolution.* Cambridge: Cambridge University Press, 1959.
Solé, Ricard, and Brian Goodwin. *Signs of Life: How Complexity Pervades Biology.* New York: Basic, 2000.
Solé, R. V., S. Rasmussen, and M. Bedau, eds. "Theme issue: Towards the artificial cell." *Phil. Trans. R. Soc. B* 362 (2007): 1725–1925.
Spencer, J., and D. Grinspoon. "Inside Enceladus." *Nature* 445 (2007): 376–377.
Summers, William C. *Félix d'Herelle and the Origins of Molecular Biology.* New Haven: Yale University Press, 1999.
Surrey, T., F. Nédélec, S. Leibler, and E. Karsenti. "Physical properties determining self-organization of motors and microtubules." *Science* 292 (2001): 1167–1171.
Surridge, C., ed. "Computational biology." *Nature* ("Insight" Supplement) 420 (2002): 205–251.
Szabo, P., I. Scheuring, T. Czaran, and E. Szathmary. "*In silico* simulations reveal that replicators with limited dispersal evolve towards higher efficiency and fidelity." *Nature* 420 (2002): 340–343.

Szostak, J. W., D. P. Bartel, and P. L. Luisi. "Synthesizing life." *Nature* 409 (2001): 387–390.

Thomas, D. N., and G. S. Dieckmann. "Antarctic sea ice—a habitat for extremophiles." *Science* 295 (2005): 641–644.

Thomson, W. Presidential Address to the British Association for the Advancement of Science. *Nature* 4 (1871): 262.

Tice, M. M., and D. R. Lowe. "Photosynthetic microbial mats in the 3,416-Myr-old ocean." *Nature* 431 (2004): 549–552.

Tort, Patrick, ed. *Dictionnaire du darwinisme et de l'évolution.* Paris: Presses Universitaire de France, 1996.

Troland, L. T. "Biological enigmas and the theory of enzyme action." *The American Naturalist* 51 (1917): 321–350.

Tuerk, C., and L. Gold. "Systematic evolution of ligands by exponential enrichment: RNA ligands to bacteriophage T4 DNA polymerase." *Science* 249 (1990): 505–510.

Ueno, Y., K. Yamada, N. Yoshida, et al. "Evidence from fluid inclusions for microbial methanogenesis in the early Archaean era." *Nature* 440 (2006): 516–519.

Van Zuilen, M. A., A. Lepland, and G. Arrhenius. "Reassessing the evidence for the earliest traces of life." *Nature* 418 (2002): 627–630.

Varela, Francisco J. *Principles of Biological Autonomy.* New York: Elsevier, 1979.

Varela, F. J., H. R. Maturana, and R. Uribe. "Autopoeisis: The organization of living systems, its characterization and a model." *BioSystems* 5 (1974): 187–196.

Vestigian, K., C. Woese, and N. Goldenfeld. "Collective evolution and the genetic code." *Proc. Natl. Acad. Sci. USA* 103 (2006): 10696–10701.

von Neumann, John. "The General and Logical Theory of Automata." Reprinted in *Cerebral Mechanisms in Behavior,* ed. Lloyd A. Jeffress, 1–41. New York: John Wiley and Sons, 1951.

Waal, Frans de. *Chimpanzee Politics: Power and Sex Among Apes.* Rev. ed. Baltimore: Johns Hopkins University Press, 1998.

Wächterhäuser, G. "Evolution of the first metabolic cycles." *Proc. Natl. Acad. Sci. USA* 87 (1990): 200–204.

Wagner, Andreas. *Robustness and Evolvability in Living Systems.* Princeton: Princeton University Press, 2005.

Wang, L., A. Brock, B. Herberich, et al. "Expanding the genetic code of *Escherichia coli.*" *Science* 292 (2001): 498–500.

Ward, Peter D., and Donald Brownlee. *Rare Earth: Why Complex Life Is So Uncommon in the Universe.* New York: Copernicus, 2000.

Ward, Mark. *Beyond Chaos: The Underlying Theory Behind Life, the Universe, and Everything.* New York: Thomas Dunne/St. Martin's, 2001.

Waters, E., M. J. Hohn, I. Ahel, et al. "The genome of *Nanoarchaeum equitans:* Insights into early archaeal evolution and derived parasitism." *Proc. Natl. Acad. Sci. USA* 100 (2003): 12984–12988.

Watson, James D. *The Double Helix.* Harmondsworth: Penguin, 1968.

Weaver, W. "Science and complexity." *American Scientist* 36 (1948): 536–544.

Webb, Stephen. *If the Universe Is Teeming with Aliens . . . Where Is Everybody? Fifty Solutions to the Fermi Paradox and the Problem of Extraterrestrial Life.* New York: Copernicus, 2002.

West-Eberhard, Mary Jane. *Developmental Plasticity and Evolution.* Oxford: Oxford University Press, 2003.

Wharton, David A. *Life at the Limits: Organisms in Extreme Environments.* Cambridge: Cambridge University Press, 2002.

Wheeler, Quentin D., and Rudolf Meier, eds. *Species Concepts and Phylogenetic Theory.* New York: Columbia University Press, 2000.

Wills, Christopher, and Jeffrey Bada. *The Spark of Life: Darwin and the Primeval Soup.* New York: Perseus, 2000.

Wimsatt, William C. "Aggregativity: Reductive heuristics for finding emergence." *Philosophy of Science* 64 (1997): S372–S384.

Winkler, W., A. Nahvi, and R. R. Breaker. "Thiamine derivatives bind messenger RNAs directly to regulate bacterial gene expression." *Nature* 419 (2002): 952–956.

Woese, C. R. "Order in the genetic code." *Proc. Natl. Acad. Sci. USA* 54 (1965): 71–75.

———. "On the evolution of the genetic code." *Proc. Natl. Acad. Sci. USA* 54 (1965): 1546–1552.

———. *The Genetic Code: The Molecular Basis for Genetic Expression.* New York: Harper and Row, 1967.

———. "The universal ancestor." *Proc. Natl. Acad. Sci. USA* 95 (1998): 6854–6859.

———. "Default taxonomy: Ernst Mayr's view of the microbial world." *Proc. Natl. Acad. Sci. USA* 95 (1998): 11043–11046.

———. "On the evolution of cells." *Proc. Natl. Acad. Sci. USA* 99 (2002): 8742–8747.

Wong, J. T.-F. "A co-evolution theory of the genetic code." *Proc. Natl. Acad. Sci. USA* 72 (1975): 1909–1912.

Yarus, M., J. G. Caporaso, and R. Knight. "Origins of the genetic code: The escaped triplet theory." *Annu. Rev. Biochem.* 74 (2005): 179–198.

Yen, A. S., S. S. Kim, M. H. Hecht, et al. "Evidence that the reactivity of the Martian soil is due to superoxide ions." *Science* 289 (2000): 1909–1912.

Ziemelis, K., and L. Allen, eds. "Complex systems." *Nature* 410 (2001): 241–284.

Zimmer, C. "Did DNA come from viruses?" *Science* 312 (2006): 870–872.

Index